Social Media and [...] Marketing for Entrepreneurs Bundle

Cost effective Facebook, LinkedIn, Instagram Marketing Strategy to build a personal brand

Facebook Marketing Tips Zero Cost Facebook Marketing Plan for Small Business

The information in the following pages is broadly considered to be a truthful and accurate account of facts and as such any inattention, use or misuse of the information in question by the reader will render any resulting actions solely under their purview. There are no scenarios in which the publisher or the original author of this work can be in any fashion deemed liable for any hardship or damages that may befall them after undertaking information described herein.

Additionally, the information in the following pages is intended only for informational purposes and should thus be thought of as universal. As befitting its nature, it is presented without assurance regarding its prolonged validity or interim quality. Trademarks that are mentioned are done without written consent and can in no way be considered an endorsement from the trademark holder.

Table of Contents

Introduction

Congratulations on downloading **Facebook Marketing Strategies: Zero Cost Facebook Marketing Plan for Small Business** and thank you for doing so. The world of Facebook marketing is growing increasingly chaotic and downloading this book is the first step you can take towards actually doing something about it. The first step is also always the easiest which is why the information you find in the following chapters is so important to take to heart as they are not concepts that can be put into action immediately. If you file them away for when they are really needed, then when the time comes to actually use them, you will be glad you did.

To that end, the following chapters will discuss the primary preparedness principals that you will need to consider if you ever hope to really be ready for a successful Facebook marketing campaign. This means you will want to consider the quality of your ads including the potential issues raised by their current costs, how they can be best utilized along with organic traffic and various reinforcements or fortifications you may need to have on hand in case of a new business opportunity.

With ads out of the way, you will then learn everything you need to know about groups and communities. Rounding out the three primary requirements for successful Facebook marketing, you will then learn about crucial page management principles and what they will mean for you.

There are plenty of books on this subject on the market, so thanks again for choosing this one! Every effort was made to ensure it is full of as much useful information as possible. Please enjoy!

Chapter 1: What Is Facebook Marketing?

Since Facebook stores data about its users which are entered spontaneously in their profile (e.g. age, location, interests), it has a very good idea of who its users are and what they are interested in. For example, just put a like on a football page to make them understand that you are following this sport.

You will understand then that the potential of subjecting these well-defined users to targeted advertisements is enormous.

In the past years, social media has become one of the most effective advertising channels, able to get new leads (contacts) and turn them into paying customers.

Facebook Ads works in both B2C and B2B and there are many cases that show an increase in results, even 5 times, after taking advantage of advertising on Facebook.

The growth of Facebook is steady; both in terms of new users and marketing opportunities, and the budgets dedicated to the AD on social media have doubled worldwide in the last 2 years, from $16 billion in 2014 to $31 billion in 2016. In 2018 analysts expect a further increase of 26.3%.

To all of this, add that the number of active users every day on Facebook was about 1.23 billion people in 2017 and on average each of these spends 50 minutes a day between Facebook, Instagram, and Messenger. With Facebook Ads, you can reach each of these users with highly customizable targets such as interests, demographics, locations, actions performed on a website, and much more.

Should I Then Do Facebook Advertising?

For how often this happens, the answer is "it depends". You have to ask yourself what your goals are and once the mechanisms of advertising on Facebook are understood, the answer will be obvious to you.

If you want to intercept an audience that might be interested in your product or service, surely Facebook Ads will be for you, allowing you to submit your ads to a specific target of users.

If you have traffic on your site or e-commerce, regardless of how it is obtained, you may want to recontact those users to repropose your ads.

Do you know that when you are looking for a product and after visiting a site, later when going to Facebook, you will find yourself the sponsored ads?

It's called retargeting. You will be able to create customized user lists based on their actions, such as visiting a particular page (article, product, landing page, shopping cart, etc.) and exploiting them for well-targeted ads.

It will be essential to have a very clear strategy for our Facebook Ads campaign. It will be appropriate to understand where the potential customer is located within our sales funnel and submit different ads depending on whether it is far or near the purchase.

We certainly cannot offer the same ad to a potential customer who does not know our brand and someone that already has our product in his chart. This is just to say that Facebook has great potential for small businesses, but needs to be used properly to get the most out of it.

Chapter 2: Disadvantages of Facebook Marketing

How much do Facebook Ads cost? How is the price established?

It is the most frequent question in this environment, but unfortunately, there is no answer other than "it depends".

On what does it depend?

From what you are advertising to who your target is, how many competitors there are and what your goals are, just to mention some aspects.

Before reaching the total costs of a Facebook campaign, it is important to understand how these costs are established. Unlike traditional advertising methods, Facebook does not have a fixed price for each placement, but follows a system of an auction among advertisers to get the ad published; this is because Facebook users can only see a limited number of ads per day.

There are various factors that determine which ad will be displayed, who will be shown and at what cost:

- Who you are targeting.

As we will see later, the possibilities of targeting are many and depending on the target you choose, you will compete with other advertisers.

- Your relevance score, engagement and click through rate.

Facebook assigns to each ad a score of relevance from 1 to 10 depending on the associated target and its response. The higher the interaction and the number of clicks, the higher your score will be and the more likely your ad will be displayed at lower costs.

- The timing.

The more competitive the sector is, the higher the costs will be. There will be periods of the year, such as during certain holidays, where costs could increase significantly. You can choose from various options, but which one is best will depend on your campaign objectives.

You can pay in various ways:

- Per impressions: pay for each display of your ad.

It is better to use this option when you want to reach as many people as possible, for example in a brand awareness campaign.

- Per click: pay only if someone clicks on your ad.

This option is preferable when we want to take users to click on a link that will take them to the site or to a specific landing page.

- Per action: pay only when a user performs a certain action, how to fill out an entry form, buy a product or visit a certain page.

With this option, Facebook will show ads to people who are more likely to take one of these actions.

So you will understand that answering the question "how much does it cost?" is complex. The important thing is to set a budget and manage it in the best way possible, reaching potential customers in a careful manner and keeping an eye on the really relevant parameters, such as cost per conversion. Not knowing what you will spend is one of the main disadvantages of Facebook marketing, but still, it is something that can be overcome easily.

AdWords vs. Facebook

This is a question that they often ask me. The answer is always "it depends".

We need to understand what the goals are that we want to achieve with our advertising campaign. It is often fundamental to combine both strategies, it all depends on the type of question: whether this is latent or conscious (or both).

If the goal is to make branding and then stimulate users who do not know us and may be interested (latent demand), the best choice is Facebook Ads, which will allow you, as we will see later, to reach potential customers. You can do this with various types of targeted campaigns and get leads.

Similarly, you can also take advantage of ads on the Google Display Network to reach potential customers by submitting your banners to specific placements.

If users are already looking for your product or service, the right approach is to use Google AdWords by creating ads on the Search Network. In this case, the user is already in a much lower part of the funnel, therefore more inclined to purchase as he is looking for your product/service.

Obviously, in most cases, we will find both types of demand and we will have to work on both platforms jointly.

The key thing is to understand where the user is inside our sales funnel and act accordingly; we will never tire of stressing it.

Going inside, it will be useful to retarget users who have shown interest in the product or service working with both Facebook Ads and AdWords through ads on the display network.

Attention: what if we intercept potential customers on Facebook through FB Ads and these then look for us on Google but we are not positioned in an organic way (without paying) for that keyword?

Simple, they will click on a competition link. The risk is, therefore, to practically advertise competitors. Understand well then how important it is, in the absence of organic positioning with SEO, that we must also have Google ads on the search network to cover some keywords as well.

The fact is that Facebook marketing is not so powerful if done alone, is something that some people see as a disadvantage.

Chapter 3: Facebook Marketing for Small Business

Small and medium-sized businesses can use Facebook marketing strategies with high margins of success. In fact, with more than 2 billion active users every month, it is impossible to remove the blue social from your web marketing plan.

What Could Be the Goals of Facebook Marketing Referring to SMEs?

Brand Awareness

Facebook is a very important tool that enables small and medium-sized enterprises (SMEs) to make their products and services known, while at the same time cultivating a very direct relationship with interested users.

If on the one hand your community, made up of people who already know your products, can follow us on the blue social, on the other side it is possible to reach people who do not know us, through spontaneous sharing, or through sponsored ones. The latter, through the creation of the right audience, allow reaching to new people potentially interested in our products.

Customer Care

Facebook is also one of the websites that best lend themselves to customer care, which is assistance to its customers. Indeed, given the announcement of future updates of the algorithm of the views of the News Feed, focus on customer care could also prove successful in terms of awarding the content posted.

Promotional content can still be valid, but using your own social page as a place to solve problems and perplexities of its users can be a key to use highly because it is able to trigger conversations between friends, debates, and an engagement appreciated by Facebook algorithms.

Direct Sales

Facebook can also be used to sell your products or services. Like an e-commerce site, the platform lends itself to the possibility of direct purchase from the page, with huge benefits for users. For small and medium-sized businesses, this opportunity is an important resource for saving resources that would otherwise have to be spent on the creation and management of an entire site.

Obviously it must be said that those who hold an important business cannot simply rely on the social network of Zuckerberg to market their products online, but it is a fact that not a month passes in which the Menlo Park team does not make availability of some new function that favors those who want to sell via the web.

How to Create a Facebook Marketing Strategy That Works?

Before starting to take action on Facebook, it is good for small and medium-sized businesses to devote time to creating a well-designed communication plan.

First of all, the goals of the strategies to be put in place must be defined.

Secondly, the public must be identified, that is the buyer, the typical customer, tracing a sort of identikit of its main characteristics such as age, place of residence, interests, level of education and more (see chapter 6 for more information about this topic).

As for the content, it will be good to dedicate only 20% of them to the promotion of your products or services so as not to tire the user with continuous offers and hype.

Finally, the tools for checking the results must not be forgotten, with the choice of the most appropriate metrics to follow in order to understand the effectiveness of the steps taken along the road to achieving the designated objectives. For example, if the setting up of a valid customer care campaign has been done, one of the ways to evaluate the effectiveness of the actions carried out is the analysis of the number and quality of comments received, rather than that of 'likes'.

Facebook for Small Businesses

Why should you use Facebook to market your small business?

Facebook is about to touch the ceiling of a billion users according to the latest official data released in July. The people who connect to this social network are 955 every month and 552 million every day; more than half a billion, even the number of monthly users who connect to Facebook with a mobile device - mobile phone, smartphone, tablet.

Facebook is becoming, for many, a major source of information more and more often, instead of connecting to the homepage of newspapers to see what is happening. We scroll the dashboard reading and commenting on the news linked by our friends. The 2011 CENSIS report on the American company shows that Facebook is used as a source of information by 26.8% of Americans, a percentage that grows to 61.5% in the age group of 14 to 29 years.

The quantity and nature of our relationships have been radically changed by the possibility of keeping in touch with people we do not see daily in person, but for the most diverse circumstances, we feel close. They were our friends in the past, we shared a travel experience or study, or we met online and subsequently met live.

This allows us to listen and exchange opinions, information, points of view, emotions; in this impetuous dynamic of conversations. This is an always open bar where people pass from one group to another participating in dozens of discussions. The companies suddenly find themselves "degraded" to one voice among others, which must gain attention thanks to the importance of what it says without the possibility of massively occupying the spaces of visibility. Furthermore, we must learn to speak "with" people, which mean, first of all, to listen and respond.

Should your company/association/electoral committee/theater company/excitement have a presence on Facebook? In the vast majority of cases, the answer is yes. Do not fall into the trap of thinking that Facebook is exclusively the realm of lazy people; often times it is a great way to "feel the pulse" of your stakeholders and can intercept your needs and opportunities that otherwise you would not have known.

In addition to your site where people go once in a while, or maybe they never go back; on the contrary, many of them open Facebook every day, several times a day; and, if they find what you publish to be interesting enough for them to click on "like" or leave a comment, this makes you visible on the bulletin board of their friends, not in the anonymous way of a flyer tucked in the mailbox, but with the social support of word of mouth.

An effective presence on Facebook can help you:

- Increase your visibility, spreading the posts of your blog, the videos you shoot, and the photos you take.

- Establish a more intense relationship with your customers, better knowing their needs and obtaining important feedback on what you do.

- Motivate and gratify your "super fans".

- Promote and share initiatives, special offers, and new products.

Facebook Marketing Ads for Small Businesses

Each Facebook campaign consists of 3 levels and it starts from the campaign level, which consists of one or more ad groups.

As you have just read, for each campaign you create, you will have to choose a goal. This is the real distinctive factor at the campaign level.

At the Ad Group level (Ad Set), you will have to choose the target, the available budget, the publication times, the offer and the placements (placements).

Going down the hierarchy, at the level of the announcements, you can set the type of announcement (image, video, carousel, etc.), all the texts, the call to action (action button) and the destination links.

As mentioned, the structure is hierarchical, so if you pause (or delete) for example a group of ads, the same thing will happen to all ads below that group.

The Definition of the Goals

Now that we understand the structure of a Facebook campaign and what are the parameters to be set for each level, we are ready to launch our first campaign.

The first question is, therefore "What is the goal to be achieved?"

Do you want to sell a certain product, because maybe you have an e-commerce store, want to create awareness or reputation, do you want to have leads or what?

Often, in a complete web marketing strategy, we will have to create different campaigns for the different phases of the purchasing process. We can then create different ads depending on whether the target user does not know our brand, or knows it but does not know our product/service, or for example, knows our product/service and may be interested in a commercial offer.

Facebook itself in the creation phase will propose you different objectives divided into 4 macro-categories; let's see them in detail one by one.

Brand Awareness

When to use it: in large-scale campaigns, when there is not a particular action that you want to take to the user. This goal will be more attractive to large companies that can afford to launch campaigns for pure branding. For smaller companies, however, almost every other objectives will give better and more significant results.

Reach

When to use it: similar to the brand awareness goal, the reach objective is functional to reach the maximum number of users to which the ad will show. With the introduction of the rules, Facebook now allows you to put a cap on the frequency with which the ad is shown to the same user; in this sense, the goal for reach becomes very useful when you have to work with a relatively small audience and you want everyone to view the ad.

Traffic

When to use it: when we want to take users to a website, or for example on a landing page. It is a very interesting goal when promoting content, such as a blog post.

Leads

When to use it: the lead ads greatly simplify the signup process from mobile devices. When someone clicks on the ad, a form opens with all personal contact information already pre-filled based on the information they share on Facebook, such as name, surname, phone and email address. This aspect makes the process really fast and within 2 clicks, one to open the ad and one to send the information.

The only problem with this type of objective is that often the email address used to sign up for Facebook several years ago is obsolete and has not been updated for too long. In this case, we would get a useless contact. As a result, it has been seen that better conversion campaigns perform that point to specific external landing pages with data to be filled out.

Another aspect to keep in mind is that lead ads do not allow you to include all the information you want in the offer, like on a landing page. Therefore, for campaigns that require a great deal of cognitive attention from the user, a campaign for conversions will be more successful.

That said, in any case, it is always better to do a test between the two approaches and see which performs better, because each case and sector can behave differently.

Chapter 4: How to Use Facebook Groups

A Facebook group is a micro-community within the largest community of the network that, focusing on specific themes, attracts people to the target.

To this, we add that the groups have recently had the blessing of Facebook after the hard blow to the business pages in the latest updates of the EdgeRank algorithm.

What happened to the Facebook algorithm?

The turning point came with the announcement of Zuckerberg and the subsequent official confirmation on the blog of Facebook news. In fact, the Newsroom informs users of an epochal change: less visibility on the pages in favor of a more personal communication.

More visibility for the posts of friends and family; but also explicitly mentioned by Zuckerberg, to the groups.

Unlike pages, in fact, groups have a greater predisposition to the generation of discussions and not to the simple unilateral posting of content. Precisely this content with little interaction is the one against which Facebook wants to fight. The decision comes after the progressive descent of the engagement on the posts that are over the years has characterized the Social Network of Menlo Park.

What does this mean?

It simply means that the posts on the groups will have greater visibility and the possibility to reach more easily your target audience. It seems to me, therefore, an excellent idea not to neglect them.

How to Find Facebook Groups in the Target

Let us imagine, of course, that you have already identified the niche or sector on which to base your communication strategy. The first of the problems you need to solve is to find groups to distribute your content on. I show you a couple of techniques you can use.

- Facebook Searches

If you have an active Facebook account in America, you can use the search bar to find groups that meet your requirements. For example, if you look for a group that deals with social, you can write "social media" on the search bar and then apply the filter "groups" (Facebook will initially search among posts, people and so on).

If you want to get even smarter, I recommend you to use Facebook in English. You will open other very interesting possibilities! Facebook in English supports the "Graph Search", thanks to which it is possible to create more complex searches such as: "Groups joined by my friends who like social media". That is, look for groups to which were added by part of my friends who like social media.

Or you might still be looking for users who like certain pages or groups in an area of interest. For example: "Friends who like Wired". You can then go to peek at the groups in which your friends have entered, going directly to their profile (privacy permitting!).

- Use Facebook Tips

Of one thing we can be extremely sure of is that Facebook knows us more than we think.

It is for this reason that it will be easy for them to suggest groups that might be interesting to you based on the past history of your likes and the groups in which you are already inside.

Just go to the Facebook page dedicated to groups. On the upper part of the drop-down menu, you can enter the "Discover" tab. Here you will find lots of suggestions divided by categories, including the "local" one to find geographically close groups.

This feature is available on both desktop and mobile. In the latter case, you will go to the main menu> Groups> Suggested.

I remind you that the specific features of the groups were recently incorporated into the main app of Facebook, while the app "Facebook Groups" retired in September 2017.

What to Do Before Posting Content in Groups

I already know that you're cheating to publish your content on the groups of your interest, but before starting to spam I suggest you follow some rules.

- Always look carefully at group usage policies.

Any self-respecting group manager sets up rules in their own group to prevent the occurrence of phenomena such as SPAM, excesses of OT (Off Topic, i.e. ending off the arguments to which the group is dedicated), too much promotion, trolling and so on.

This means that there may be rules on the amount of content that can be submitted to the community, for example. Or that you can publish external links to Facebook only with the permission of the admin.

Policies are often indicated in the first post above, often attached with a "pin". Or they could be in the "file" section of the group. Study these rules well and behave accordingly.

- Keep to general rules of good behavior.

Apart from the policies, which rightly dictate contents, times and rhythms, there are also rules of good common life that should always be respected. If you produce content in large quantities:

- Do not spam it all the time.

- Always check that it is in theme with the group, that they are of quality and that they can be useful and relevant.

- If possible, add a note to the article, a textual status that for example highlights a sentence of the article or that poses a question capable of generating a valuable discussion.

Remember, these are not only your contents! Subscribe to the groups, be in silent mode, and then spam your news as if there is no tomorrow, these will certainly irritate the admin that at some point they might even think to kick you out!

Even more so later in this historical moment in which the use statistics of the group have been made available to the administrators.

My advice is to always balance your content with content from other sources. And it would be desirable to participate in the discussions, in addition to the normal posting!

If you already produce your content and think that it is sufficient for your strategy, it means that you skipped the previous paragraph. Go read it!

For a content strategy, whether linked to the development of your business or to your personal branding, you need news and content about your industry.

I know I'm repetitive, but I'll write it again: you cannot just talk about yourself! To find valid content on the web, you can use many tools like the simplest monitoring software (I think of Google Alerts or Talkwalker) or news aggregator sites like Feedly.

In the era of the so-called "content shock", where our ability to absorb information is definitely lower than the amount of content produced, the figure of the creator becomes of primary importance. Skimming the contents and proposing the best to a community, therefore, becomes a primary role.

Additional Tips

Create a Network for Your Loyal Fans

This group is made up of the most loyal fans, people who believe in your company, in your product, and in your values. Brands like Canva, who has a loyal fan base, called these groups "community groups" or "ambassador groups".

The one with the loyal fans is the most important group you can create on Facebook. It is important to create a positive word of mouth on your products, your company, and your activities. Be sure to make your fans feel special. You can do it in various ways; by sending a shirt or by putting a comment on them.

The most important thing is to thank them, heartily and regularly. A written note (maybe by hand), is much more pleasant than a gift. Thank your fans for helping you grow your community and achieve your goals.

Here are 5 ways you can use these groups:

- Get feedback on new products.

You have a new product to launch but first, you want to know how it would be accepted? This is the right place to test it.

- Attract new members.

If your company has a membership program, this is a great way to keep in touch with them and attract new ones.

- Answer the questions.

How many times have you seen a question on a group or an online forum and have you thought that your product could be the ideal solution? Of course, you could answer the question with your private profile, but you will be more successful by responding to the members of your group.

Answer often to questions and to avoid spam (especially if there are many members in the group), tag two or three experts in the response that can respond appropriately.

- Recruit experts.

Many people like to feel expert and be taken into consideration. And many like to share their knowledge with other people. Experienced people can help new consumers get closer to your product and buy it. Please note that it may be necessary to create a dedicated expert group.

- Share the company's achievements.

If the company wins a prize or is named in an important publication, let the whole world know about it. You can then call your friends, family, post the news on your private profile, but in this way, you will not go very far. If your company receives a prize, share the news with the fan group.

That's why Facebook exists.

Help Your Customers

The second type of Facebook groups is dedicated to acquired customers. To build this community, invite people to join the group as soon as they make a purchase. Make customers aware of the existence of the group and that you would like them if they too registered.

Explain how the group works in the "information" section, as Golden Tote does. For example, members can get to know each other, share ideas and strategies and help each other in any way.

Be sure to send an email with the link to subscribe to the group and monitor who signs up or not. Do not forget to invite people who have not yet registered several times.

Ways to develop your group on Facebook:

- Be generous and promote a "give" policy.

For example, offer tips or tricks to use the less known features of your product and invite other users to share their findings.

You could invite members of your fan group to share tips and tricks with them too.

- Be transparent.

Respond personally to customer criticism. For example, acknowledge the customer who makes you notice what is lacking in your product and responds that you will immediately fix it.

It is not necessary to promise discounts or promotions to the disgruntled customer, but it is important to thank him for the report, to apologize and to promise that he will be contacted when the problem is solved.

- Be present.

Answer questions and comments as soon as possible

If you cannot give an immediate answer, still let your customer know that you will respond quickly. From suggestions, tips, tricks. Encourage other users to share their findings.

- Be inclusive.

Offer group members promotions and offers that are not on the official page. Make your customers feel part of an exclusive club. If a social media user shows a particular interest in your product, he also invites you to join the online group.

Grow Target Segments (With Multiple Facebook Groups)

One of the most important goals of marketers is to develop a "typical customer", that is to identify their customers and understand what their doubts, problems and how your product can meet its needs.

If you have different types of customers, create more groups on Facebook. For example, you could create a group based on the language spoken by its users. Maybe not all customers will want to be considered experts and to answer questions from other users; you can then create smaller groups to meet the needs of each client.

Chapter 5: How to Start a Successful Facebook Group

Maybe you do not know, but creating a group on Facebook allows you to share photos, videos and anything else just with whoever you want. Thanks to groups, you can interact with a small circle of people without the others seeing what you're sharing with them.

Even if you do not consider yourself a social network expert, you can feel comfortable. In the next paragraphs, you will find all the information you need to create a Facebook group, whether you intend to act on your PC or your mobile device. In addition, you can find some useful tips for creating group chats on Messenger and use them to share various types of information with friends and relatives. What do you say? Are you ready to start? So let's put aside the talk and get straight into the heart of this tutorial. I wish you good reading and have fun!

Wanting to create a group on Facebook but unfamiliar with the social network founded by Mark Zuckerberg, you do not have the faintest idea how to proceed? No fear! Just follow the steps listed below to complete the task.

PC

To create a group on Facebook, from your PC, connect by the browser to the website of the social network (Of course, after you have already logged-in to your account.) and click on the green Create Group button located at the top right.

In the box that opens, type the name you want to assign to the group in the text field located under the entry 'give a name to the group', select the users you want to include in the latter by typing their names or their email addresses in the field text placed under the entry 'add some people' and then select one of the privacy settings available to decide who will be able to view the content posted. Select 'public group' to allow all users of Facebook to view the group and the content posted in it; 'group closed' to allow everyone to find the group and see the members but without being able to view the posts or 'secret group' to allow only members of the group to find it and view its posts.

After choosing the name of the group, the members that will be part of it and the privacy settings, click on the blue button 'create', select an icon that identifies the latter (e.g. the icon of the

46

basketball, if your new group is dedicated to the world of basketball or the camera icon, if your group talks about photography) and click on the blue 'OK button'. If you wish, you can also skip this step by clicking on the 'skip' button.

Once the procedure is complete, you will be sent back to the main page of your group, which you can customize by inserting a cover image. To do so, click on the green 'upload photo' button and upload an image saved on your PC, or click on the 'select photos' button to use an image from those you have already uploaded to your Facebook profile.

To interact with friends who are part of the group, click on 'post to write a new post', click on the item 'add photo/video' to insert a multimedia content or on the voice 'live video' to start a live broadcast with other members of the group.

After creating the group, click on the options that are located immediately below the name of the group (top left); "Information", to add a description to the group; "Discussion", to share content with other members; "Members", to view members who are part of the group and to modify their roles; "Events", to create an event on the group or "Manage the group", to create scheduled posts, accept subscription requests, view the activities of administrators and so on.

To change other settings concerning the new Facebook group you have created, click the 'more' button (after clicking on the 'discussion' option) and, in the menu that appears, click on one of the options you see: "Add members", to add new members to the group or "Edit group settings", to change their name, select the group type (that is, the category related to the topic dealt with in the same), link a page, select who can publish a post and so on.

Android

To create a group on Facebook from your Android device, first, start the official social network app on your smartphone or tablet and log in to your account (if required). Then tap on the symbol (≡) located at the top right, press the 'create group' button (located under the heading Groups), type the name of the group in the text field 'give a name to this group' and, if you wish, press on the camera symbol to take or add a photo.

Then add the members to be included in the group by selecting them from the menu located at the bottom of the screen and press the 'next' button. At this point, select the privacy settings by ticking one of the available options: "Public", to create a public group and then allow anyone to view the posts and members that are part of it; "Closed", to hide the posts published on the group to those who are not members or "Private", to make the group "invisible" to those who are not part of it. After indicating the privacy options you prefer the most, press on the item 'create'. Et voilà! You've just created your group on Facebook!

Even on Android you can add new members, edit group info, create events, post photos, files and anything else using the appropriate commands that you find immediately under the name that identifies the group.

iOS

To create a group on Facebook from your iOS device, start the Facebook application, log in to your account (if necessary), press the symbol (≡) located at the bottom right and then tap on the item "Groups". In the screen that opens, press the symbol (+) located at the top right, type the name of the group in the text field "Give a name to this group" and, if you wish, press the symbol of the camera to shoot or add a photo to be used as a cover for the group.

Now, choose the members to add to the group by selecting them from the menu at the bottom and press the 'next' button. Then select the privacy settings by ticking one of the available options: "Public", to create a public group; "Closed", to hide the posts published on the group to those who are not part of it or "Private", to not allow those who are not members of the group to find it, nor to display the posts. Finally, press the item "Create" to complete the group creation procedure.

Now you can finally customize your new group, add new members, edit the info, post photos and anything else using the buttons that are immediately below the name and photo of the group itself.

Create a Chat Group Along with Your Group

If your intention is simply to converse with other Facebook users, you probably do not need to create real Facebook groups, but you can simply create simple group chats using the appropriate feature included in Messenger.

- To create a group chat on PC, log in to your Facebook account from the browser, click on the "Messenger" icon (the lightning bolt symbol at the top), click on the item "New group", type the name of the group chat in the text field "Assign a name to the group", select the users to add to the chat and click the blue "Create" button.

- To create a group chat on smartphones and tablets, however, start the Messenger app on your device, press on the voice "Groups", tap on the item "New group" (on Android) or "Create" (on iOS) and finally press on 'arrow symbol' (on Android) or on the "Create group entry" (on iOS) to complete the operation.

Chapter 6: How to Manage and Grow Your Facebook Group

Although it may seem a lot less "cool" than Facebook Fan Pages, a successful Facebook group can be a very interesting channel for acquiring web traffic.

In this chapter, we will discuss some very useful tips to quickly grow a Facebook group.

Before entering into the heart of this topic, I would like to make a small premise.

As you may already know, a new business begins to achieve extraordinary results when it manages to create around itself a community of similar people with shared interests.

By dealing with digital projects, our focus will be moving towards a specific type of community: the online community, the concept that is right behind a Facebook group.

Anatomy of an Online Community

Generally, an online community consists of three distinct categories of users:

- 90% are lurkers in public;

- 9% is the average contributors;

- 1% is loyalists.

By lurker, we mean all those users, who hide among the members of a community, and who observe, read and use the information shared by the group, but without then going to interact and feed the conversation.

Why do they behave this way? Maybe because they believe they are not up to certain discussions, perhaps because they are shy people even in digital, or because an active participation still entails a waste of energy and time to devote to the online conversation.

Then there is the average contributor, those who together make up 9% of the community, and who share useful contents from time to time, with a variable frequency.

Lastly, there is the hardcore of the online community, 1% of the "Highlander", or rather the loyalists. They are the ones who check every day the postings spread in the group, ask questions provide feedback to other users, and that would continue to feed the conversation in the group if it were not that at some point of the day you still have to eat and sleep!

Why Should We Focus on the Online Community of a Facebook Group?

One of the greatest benefits of developing a community on Facebook is the concept of target reach.

It is well known that a post published on a Facebook page organically reaches only a very small part of the fans of that page while, if you post the same content on a Facebook group, such sharing will be notified to every single member of the community.

So, the coverage of a post is far higher within a Facebook group and therefore also all actions related to it.

It is a very useful tool to generate web traffic for a company or personal site. But to be really effective it must be based on interesting metrics; the first of all is certainly the number of participating members.

I would, therefore, like to share some very useful tips to quickly grow the user base of a successful Facebook group.

- Perform a preliminary analysis.

First, you need to ask yourself what the topics are, and these will be the subject of discussion of the group you want to develop.

As a result, you will be able to outline a profile of your ideal target. You cannot aim to have anyone enrolled in your Facebook community; you have to aim for the right people.

Ask yourself what your group is about. This will also be the answer concerning who really is your ideal target and what are its main characteristics.

For more profiling, you can always make sure that the members themselves define themselves better. Create market research in which you try to extrapolate more or less information about the people themselves.

- Set goals.

Once you have a clear idea of the context on which your group revolves, it's time to think about what goal you want to pursue thanks to the Facebook group.

Do you do it for personal interests than for your brand? Why do you like networking with professionals like you? Or why do you simply want to create a community of people available to each other?

Whatever your goal, having defined it will help you adapt each situation to pursue the search results.

- Set it up the right way.

A successful Facebook group cannot ignore, first of all, an effective name, which is as simple as it is attractive, and which can arouse curiosity and interest.

Also, think about any keywords that may be useful for searches on the internal Facebook engine.

About keywords, on the sidebar of a Facebook group or in the general settings (which you can access by clicking on the three points on the top right, near the cover image), you find a field where you can define up to 3 tags, useful for being able to be found by people.

For proper setup, take into consideration the clear description that explains what topics are treated in the group and what kind of people are registered.

In addition, you will also have to decide whether to set the group as closed, public or secret.

Starting from the last just mentioned, a secret group can be found on Facebook only by its members while a public group allows anyone, even those who are not part of the community, to see the posts shared within it.

The most effective type to quickly acquire new members in the Facebook group is the closed one.

A closed group can be found by any user during a search on Facebook using the appropriate keywords but only members can see the content published in it.

Therefore, those who are really interested in consulting what is within the group must first register. This could be an interesting profiling process though, so think about it before deciding whether to create a public or secret group.

- Define the rules.

A healthy relationship, even the virtual one, between people must be based on clear and precise rules.

Without the guidelines, anarchy would reign, and anyone could do what they like. For example, you could only share links related to your site, constantly spamming, and you could respond in the comments using unpolished tones.

This could cause serious damage to the image of the group, difficult to repair in a short time.

In addition to this set of rules, which in digital jargon is called netiquette, you must also know how to handle cases in which heated discussions are developed between the various members.

We must, therefore, intervene, bring order and try to make sure that members can return to converse with calm tones. It is the so-called crisis management.

If, in the case, certain unpleasant situations persist, such as the continuous sharing of posts related to personal services, these articles should be removed. The responsible person should be warned and possibly banned, should he repeat this behavior several times.

- Share useful content.

Doing Content Marketing is essential everywhere on the web, let alone in a successful Facebook group.

If you do not share valuable content, you do not feed the discussion among the members, and in that absence, there are no reasons for having to be part of an online community.

So, it is essential to design an editorial plan for your Facebook group, both in terms of content creation (creation of original content) and content curation (sharing of useful material created by other people).

In addition, if you want to optimize your work, you can always use one of the various digital tools available for post-programming on a Facebook group. For example, you could try PostPickr; it is a very useful tool that can help you take your business to the next level.

To increase the acquisition of new members and above all increase the engagement of those already present, as the group's admin, you have to publish with a certain regularity, perhaps even including questions in the texts of the posts to fuel a constructive debate between the people.

It could be once a day, twice a day, or even less, such as 2-3 times a week.

The important thing is that you understand one thing; you have to offer quality content that can really arouse interest and be useful to your subscribers.

Also, remember two other interesting features of Facebook groups:

- You can highlight a post compared to the others (the "Pin the post" feature). For example, if you find that a content can prove to be of absolute value for all members or if you want to keep the rules of the community easy to consult.

- In addition to publishing links, you can also upload files (for example, a pdf file), conduct surveys, direct live events, organize events and create documents linked to the group.

- Shape your team.

In the beginning, when the group consists of a few tens or hundreds of members, it will not be necessary to have other collaborators to manage the community.

But when you start having thousands of participants, the situation could get out of hand; too many requests for registration links sharing of dubious usefulness, colored verbal exchanges between members, etc.

A simple remedy is to ask for help from other people, in short, set up a team of admin and moderators of the group.

Both these figures can:

- Approve or deny access to the group.

- Approve or deny the publication of a post.

- Remove posts and comments.

- Remove and ban members.

- Highlight a post.

The substantial difference that exists between an administrator of a group and a moderator is that the former can offer and remove the role of admin or moderator to other members and can make changes to the general settings of the group (name, cover photo, privacy level, etc.).

- Promote the group.

If your tactics for organic and natural growth are tight and you want to take advantage of other techniques to quickly increase the number of members of your community, I suggest you try the following strategies.

First of all, you could start Facebook advertising campaigns to reach new people potentially interested in your group.

The Facebook Ads do not allow you to create a post natively sponsored for the promotion of a group but there is a simple trick to turn the obstacle.

So what you can do is create a Facebook page related to your group, create an advertisement linked to it and enter the URL address of your Facebook group in the field associated with the website to be sponsored.

In this way people will be conveyed directly to your online community and, if they are really interested in the topics you have proposed, they will not hesitate to register.

Another possibility to increase the number of members is represented by cross-promotions. It is about making new collaborations with other Facebook groups more or less related to the issues faced by your community. The purpose of this collaboration is based on the exchange of promotions between groups; just publish the URL address of the partner group and invite their members to follow the other group, if they consider it appropriate.

Another good way of promoting your groups is to mention the group in forums like Redditt and Quora on questions related to your niche. Just answer the questions and mention at the bottom that you have a free Facebook group that anyone can join.

Chapter 7: How to Design Your Customer Avatar

To define the profile of your ideal customer you have to remember that each individual is influenced by his position in society. Therefore, to trace the profile of the ideal customer we must answer questions related to these influences.

Social Influences

Cultural Systems and Subsystems

How old is he?

Where does he live?

Is he a man, a woman, or both of them?

What degree of education does he/she have? What schools did he/she attend?

What hobbies or passions do they have?

What are its values?

What are their beliefs?

Social class

What kind of work do they do?

How much does he or she earn?

Family

Is he or she married?

Does he or she have children? How many?

Does he lives alone and is single?

Does he live by his parents?

Marketing Influences

Do they already use a product to meet the need or solve the problem?

What kind of product do they use?

What kind of brand?

What features are relevant?

What is the benefit most appreciated?

On what price range is it oriented?

Is the high price for the customer a way of affirming their social status?

Do they associate high price with quality?

What kind of advertising influences their purchase?

Where do they usually buy?

Situational Influences

Which environments can influence the purchase (showroom, the point of sale)?

Social Environment

Within the group in which he/she lives and works, who else influences the choice of purchase?

Who uses the product?

Who pays for the product?

Are the influencer, the consumer, and the buyer the same person?

Psychological Aspects Associated with the Product

What are their fears?

What problems do they want to solve?

What consequences would it entail for an unsolved problem?

What are the challenges they're facing?

What are their wishes?

Emotional Aspects Related to the Choice of the Product

What mistakes are they afraid to commit by making a wrong choice?

What would it mean for the customer to choose a wrong product?

Now that you have answered these questions, you can trace the profile of the ideal customer.

Here's how to proceed.

Put a face to your potential client (download a photo from the internet that identifies the physical characteristics of your potential customer) and a fictitious name.

Then compile the data that will trace the characteristics of socio-demographic, psycho-graphic and consumer experiences.

This serves to have a clear representation of him/her with whom we are going to talk and with whom we want to relate.

This exercise will help you understand your customer' motivations and tailor your marketing efforts towards those motivations.

A great resource to boost this process and find people that met your avatar standards on Facebook is **lookup-id.com**. Thanks to this website, you can define your target customer through the definition of personal characteristics and get a list of people that met them.

Here is how it works.

Once that you have entered the website, go on the "extract members" section. It is very easy to find on the top right of the page. From there, you want to insert the ID of a Facebook group in your niche, which of course will contain people in target with your offer. The website will give you a complete and detailed list of the people in the group, which means that you have just discovered a goldmine, since those will be in target for what you are offering.

On this website, you can even use the FB Search function. Once you have designed the features of your typical prospect, you can insert them in this platform to get a list of users that meet those standards. It is pretty straight forward, and it is very easy to use. Our suggestion is to play around with the website and get a grasp of its potential: once you start using it, you will never get back at the classic manual research.

Now it is time to use another great tool to make life easier and start gathering a following. First of all, install **Toolkit for Facebook by PlugEx**. What is this? It is an amazing tool that will allow you to do multiple actions in a matter of seconds. For instance, once you have your list and have launched Toolkit by PlugEx, you can then invite all those people to put a like on your page or to join your group.

When you start using these tools on a regular basis, you will be amazed at how easy it is to grow your fan base and start getting significant results. One little tip that we like to give our readers is to always use a secondary account for these operations, in order to guarantee proper privacy protection.

A practical example

Now that we have seen how to use the two software in theory, it is time to dive into the practice and use a practical example to understand the concepts better.

So, let's say that we have a shop that sells running gear and we want to find customers that are in target with the items we sell. The first thing we want to do is to go on running groups, like this one https://www.facebook.com/groups/TrailAndUltraRunning/ and look for the group id. We can do that by going here https://lookup-id.com/ and entering the previous URL. This will give us the ID of the group. After that, we just need to paste the group ID here https://lookup-id.com/get_facebookid.php to get a list of all the members that are inside the group.

After having done that, you can use Toolkit for Facebook by PlugEx to quickly invite all the people on the list you just found to leave a like on your dedicated running store page. Furthermore, you can even invite them in your very own group, where you will start your marketing process.

This is how you can use the two software together to really boost your group following.

Ref: Business Guide Offer, Instagram University

Chapter 8: How to Use ManyChat

If you're not using ManyChat, right now you're losing 60% of new potential customers.

(You're literally burning your money!)

Does this seem absurd?

Perhaps you do not know that the Chatbots are the future of marketing. For now, know that they have incredible power; to enter people's lives like never before.

Of course, there are e-mails, but, think about it. Would anyone really open a promotional e-mail?

The truth is that some when they find them in the mailbox, they almost automatically discard them. They have developed a natural tendency to delete promotional emails.

Associate the email with invasive advertising on at work or, again, at the studio. To talk to a friend, however, they use WhatsApp or Facebook Messenger.

What Do We Want to Say?

You should choose the tool to use based on your audience. If, for example, it is made up of managers, who often check e-mail, e-mail marketing may be the ideal choice.

But if perhaps, you turn to students, on average 20 years, who have much more convenience with the chat, it would be better to think of a Facebook Bot.

Do you understand what the extraordinary advantage is?

It will allow you to establish an authentic communication with your (potential or not) customer, who will open your message just like a friend. You will receive a notification on your mobile phone, you can chat with you; everything automatically.

The only thing you will have to deal with is the Bot settings.

The good news is that it is an extremely simple operation. Thanks to the services available you will use more or less 10 minutes.

However, without knowing how to use this very powerful tool effectively, you would risk undermining this work. This is why in this chapter we will introduce you to the correct configuration of a Chatbot with ManyChat.

Are you ready? Let's start!

What Is ManyChat and Create a Facebook Bot in 10 Minutes

We can say that a Bot is a program that is able to manage, in an automatic and natural way, conversations in a chat with users.

It can answer questions, offer solutions to problems and make proposals, just as if it were human.

Already from this general definition, you will have guessed that you have in your hands something potentially revolutionary for you and your business. You will free your time, increasing your results.

ManyChat, What Is It?

In short, it is the simplest and most intuitive service to create a Facebook Messenger Bot.

You do not need programming knowledge. The configuration is extremely fast, and you will immediately have the opportunity to carry out a series of actions:

- Create automated message sequences.

- Send a message to all users registered in the bot.

- Use advanced tools to increase conversions.

How ManyChat Works: The 2 Basic Tools for Bot Marketing

We have just said that ManyChat is the easiest to use Messenger Marketing tool.

Despite its simplicity, however, it offers a number of features that make it complete and effective for your web marketing strategy.

Do you want some examples?

- Automatic sequences.

This is the series of messages that, automatically, ManyChat will send to the user. You can set them as you like, based on your lead generation strategy.

- Growth Tools.

Here, this is the real bomb among the tools offered. They are a series of "extensions" that add functionality to ManyChat. The most famous is Facebook Comments Tool, which allows you to convert whoever who comments on a particular post into a member of the Bot.

(These are just some of the features you'll have available.)

Well, to make you better understand the functioning of a Chatbot; let's assume a case of real use:

The user Luke comments on the post of your product with the specific keyword that you have set. Your bot turns on and automatically sends your opt-in message; basically, a welcome message to confirm the user's interest.

Subsequently, based on the behavior of Luke, the bot will send him different messages, to achieve the goals you have set. These messages are part of an "automated sequence" that you created earlier.

The important thing is to never be intrusive - keep this in mind during the setup phase. Each time you send a message, a notification will be sent to Luke's mobile phone. This means two things:

- You will enter his daily life, like never before (remember what we said at the beginning of the chapter?).

- You will have to manage this opportunity in the best way, so as not to frustrate him (most people are not accustomed to this tool just yet, so do not overuse it).

Conclusion

Thanks for making it through to the end of **Facebook Marketing Strategies: Zero Cost Facebook Marketing Plan for Small Business**, let's hope it was informative and able to provide you with all of the tools you need to achieve your goals whatever it is that they may be. Just because you've finished this book doesn't mean there is nothing left to learn on the topic. Expanding your horizons is the only way to find the mastery you seek.

The next step is to stop reading and to get started doing whatever it is that you need to do in order to ensure that those you care about will be properly taken care of should the need arise. If you find that you still need help getting started, you will likely have better results by creating a schedule that you hope to follow including strict deadlines for various parts of the tasks as well as the overall completion of your preparations.

Studies show that complex tasks that are broken down into individual pieces, including individual deadlines, have a much greater chance of being completed when compared to something that has a general need of being completed but no real timetable for doing so. Even if it seems silly, go ahead and set your own deadlines for completion, complete with indicators of success and failure. After you have successfully completed all of your required preparations you will be glad you did.

Once you have finished your initial preparations it is important to understand that they are just that, only part of a larger plan of preparation. Your best chances for overall success will come by taking the time to learn as many marketing skills as possible. Only by using your prepared status as a springboard to greater preparation will you be able to truly rest soundly knowing that you are prepared for anything and everything that the market decides to throw at you.

Finally, if you found this book useful in any way, a review on Amazon is always appreciated!

LINKEDIN MARKETING

Use LinkedIn B2B Marketing to Generate Qualified Prospects and Obtain Clients

BOOK DESCRIPTION

LinkedIn is the world's largest professional network platform. With over 500 million professionally-connected members, this is a giant opportunity for your Business-to-Business (B2B) marketing. This book guides you on how to use LinkedIn B2B marketing to generate qualified prospects and obtain clients.

To appreciate what awaits you should you take up LinkedIn advertising, this guide begins by providing you with proven benefits of LinkedIn advertising so that you may know your gains well in advance.

While LinkedIn has been used successfully by many companies to market their businesses, it is not a one-platform-fits-all kind of basket. There are those types of business that have a higher potential to gain from the uniqueness of LinkedIn platform and there are those that have a lower potential of deriving the same gain. This simply means that LinkedIn must be evaluated on a case-to-case basis. In this guide, you will find valuable criteria that will enable you to determine whether LinkedIn suits your marketing needs or not so that you do not risk your resources against potentially low ROI (return on investment).

Professionals like belonging to clubs. Online, groups are the equivalent of professional clubs. Thus, you need to know how to use LinkedIn groups to your marketing advantage. This guide will show you how you can leverage the power of groups to gain a competitive edge in your niche.

81

LinkedIn Ads are quite unique from other forms of online Ads. This means that you have to give them special attention by learning how to effectively use them for maximum results. We are going to show you how to use the various types of LinkedIn Ads to increase your income flow.

What is the benefit of having a large following without an impact on your revenue? None! Most enterprises make the mistake of working so hard to acquire a huge following on LinkedIn without knowing how to financially gain from this following. We provide you with practical approaches to grow and monetize your LinkedIn following so that you can increase your income.

Finally, there is a limit to how much you can do as an individual. Even if you employ a dozen more people to work specifically on your LinkedIn account, you cannot completely tap into the huge potential base of over 500 million people. This is where automation comes in handy. Technologies exist that can help you to effortlessly automate your LinkedIn income. This guide provides you with information on how you can automate your LinkedIn income and the specific tools you require in order to harness great success.

Kindly download or print this book to learn more.

Enjoy your reading!

DISCLAIMER

INTRODUCTION

Social media networks have become a great source of potential leads. LinkedIn is one such social media network. However, unlike other social media networks, LinkedIn is a dedicated professional network characterized by members who are successful professionals, leaders in their respective companies, industries, and fields. LinkedIn has become such a marketing goldmine for Business-to-Business (B2B) companies. You too should not be left behind in digging up this goldmine. We have prepared this guide to help you take a vantage position in mining this great opportunity.

The practical information provided in this guide shows you how to use LinkedIn B2B marketing to generate qualified prospects and obtain clients. You will learn how to grow and effectively monetize your LinkedIn following. You will also learn how to scale-up your income flow through automation so that how much you earn is not constrained by limited human resource capacity but rather advanced forward by technology.

Keep reading to learn more.

BENEFITS OF LINKEDIN ADVERTISING

LinkedIn has the highest saturation of professionals and enterprise decision-makers. This makes it unique compared to other social media networks. Thus, your advert is more likely to reach the targeted audience for business than any other network.

The following are some of the benefits of LinkedIn advertising:

- It is highly targeted

With such a high concentration of professionals, executives and business leaders, this makes the best place to expose your B2B (Business-to-Business) advert.

- It is well-suited for B2B advertising

Generally, LinkedIn is not suitable for B2C (Business-to-Consumer) kind of advertising. This is where other social media do well. However, LinkedIn is unrivaled when it comes to B2B targeting. With LinkedIn, your advert is more likely going to grab attention of the eyes of the companies' top captains than on other social networks.

- A great variety of ads to choose from

LinkedIn has a unique form of advertising compared to other social media. Its advertising features are highly personalized for a professional touch. Besides that, they are plenty of options to choose from that suit your particular needs. As we shall see later under the section "How to Use LinkedIn Ads", there are a variety of unique Ad types not found in any other social media.

- Easy connection with influencers

LinkedIn has 61 million high influencers. The biggest advantage is that most of them are professional influencers. As such, they are highly likely to connect with you if your business aligns with their professional needs. They are not the highly aggressive types that are in for monetary inducements.

- Long-form content marketing

Unlike other social media, LinkedIn provides an opportunity for long-form content marketing. Long-form content allows you to dwell deeper into your brand and make your audience have an incisive perspective. Since most of the people on the LinkedIn are professionals, they are more adept at such content as it enables them to understand your company and brand much better.

- Better SEO

According to Hubspot study, LinkedIn does well on SERP (Search Engine Results Page) placements as a network that delivers the best improvement in search rankings per activity. Thus, you are more likely to get traffic to your web page from LinkedIn activity than from any other social media activity.

- Leveraging LinkedIn groups for your adverts

Most of the highly impactful business activities take place within LinkedIn groups. It is from these groups that you are able to gain personalized professional touch. LinkedIn groups are good for content marketing. Thus, you can use your posts and discussions in LinkedIn groups to advertise your brand in the subtlest way yet experience a much better outcome.

- Reach out to high-income professionals

LinkedIn demographic has a high concentration of senior professionals who range in the 30-59-year age bracket and an average income of $75,000 per year. Not only are a significant number of them in positions capable of making B2B purchase decisions, but they too have higher purchasing power for consumption. This means that if you are dealing with high-end consumer products and services, you are more likely going to receive a better response.

- Filters for advanced targeting to industry-specific variables

LinkedIn focuses on industry-specific variables for targeting audience. Such variables include;

- Industry
- Company size
- Company name

- Skill
- Degree type and name
- Seniority
- Job function
- Job title

These parameters are highly suitable for B2B targeting. They are also good for B2C targeting when it comes to offering professional opportunities such as recruitment.

- High-level lead nurturing capabilities

LinkedIn has Lead Accelerator feature. This feature allows one to track high net-worth prospects and precisely target them. It also offers opportunities for list-based advertising and remarketing to recent web visitors. All these features help to nurture leads on their journey to becoming customers.

- Set own campaign

LinkedIn has a self-serve interface that allows one to easily setup own advertising campaign without requiring external help. This makes it easy for small-scale business without an adequate budget to higher specialists to do this on their own. Nonetheless, self-serve features are more elaborate and more advanced such that regardless of the firm size, the purpose is well served.

- Get the 'right' clicks

LinkedIn high-precision parameters make it easy to get the right high-value clicks rather than plenty of low-value clicks with a low viability of conversion. It is better to get fewer clicks that convert than so many clicks that hardly convert. If your campaign is based on CPC, then, this guarantees you high ROI on your investment.

- Higher conversion rates

With the right audience, advanced filtering, specialized lead nurturing and focus on the right clicks, you are guaranteed higher conversion rates compared to other social media networks.

IS LINKEDIN ADVERTISING RIGHT FOR YOU?

As we have stated earlier, although LinkedIn can be used for B2C targeting, it is highly suitable for B2B targeting. That should be your first criteria for determining whether LinkedIn is right for you or not. However, there are always exceptions to any rule. For example, if you are rendering professional services, executive training services, car loans, mortgages, and such other services and products that professionals would love to buy, then, B2C can work for you. Nonetheless, the B2C products and services to offer are quite narrow.

To understand if LinkedIn Advertising is right for you, let us have a look at what the statistics are saying.

Info statistics:

- 94% of B2B Marketers use LinkedIn to distribute content. [Source: LinkedIn Business]

- 80% of B2B leads come from LinkedIn. [Source: LinkedIn Business]

- 79% of marketers consider LinkedIn as a potential source of B2B lead generation. [Source: LinkedIn Business]

- Over 70% of LinkedIn members are outside the United States. [Source: LinkedIn Business]
- 57% of LinkedIn visitors access it via mobile phones. [Source: LinkedIn Business]
- 46% of social media traffic to B2B Company sites comes from LinkedIn. [Source: LinkedIn Business]
- 43% of marketers claim to have sourced a customer from LinkedIn. [Source: LinkedIn Business]
- Before a buy decision is made, decision-makers consume about 10 pieces of content from a given source. [Source: LinkedIn Business]

From the statistics, you can see clearly that LinkedIn is more cut out for B2B endeavors than B2C endeavors. With 80% of B2B traffic flowing from LinkedIn which has just about 100 million active members per month compared to Facebook that has about 1 billion active members per month, not forgetting Twitter, Instagram, Snapchat and others highly popular social media networks combined, it simply means that LinkedIn is highly focused on B2B.

If you are in the B2B sector, then, this is a full endorsement. However, you will need to consider several other factors before conclusively establishing that LinkedIn is where you ought to be.

Such factors include:

- Target customers
- Target audience
- Competitor activity
- Budget

Target customers

Though there are more professionals on LinkedIn than other social media networks, the target customers are mostly enterprises. However, these enterprises can only be represented by individuals; the professionals, CEOs, and other decision-makers. In fact, 41% of Fortune 500 CEOs are on LinkedIn. Thus, if you are targeting one or several of the Fortune 500 companies, LinkedIn is the right place to find their representatives.

Target audience

Targeting audience on LinkedIn is not a wild-goose chase. Thus, you not only need to target the right audience but also must be knowledgeable about them. The more precise you are about your target audience the higher the ROI you can derive from each target.

If you are targeting highly educated professionals, then LinkedIn is the right platform for you. Knowing the following will guide you in refining your target:

- Know your target audience's job titles. This is important if you are offering customized service to an audience within a specific title. For example, if your organization is to facilitate audit assistants to become certified, then, you will filter out to get all those who have the title 'Audit Assistants' and target your ads to them.

- Know the workplace of your target audience. There are occasions where you want to target a certain corporation, yet you do not have direct connections. The best way is to use LinkedIn filters to get employees working for that company and target them with your ads. This could be for job hunting or probably advertising certain equipment that can suit their needs.

- Know the roles played by your target audience. This is important when it comes to decision-making. The more senior the target audience, the more likely they are going to influence decisions regarding your advert.

- Know the unique skills and interests of your target audience. Knowing your target audience skills and interests is extremely important. These are things that they hold close to their hearts. Having a product or service that requires a particular skill to handle or advances a particular skill or interest will more likely result in a better response to your ad.

Competitor activity

If you are in the B2B sector, know that your competitors are more likely going to be on LinkedIn than any other social media. This is supported by the statistics, which shows that 94% of B2B marketers use LinkedIn to distribute their content. Why not you? Furthermore, 79% of marketers view LinkedIn as the best place to source B2B leads.

If you do not want to be left behind by your competitors or you would like to be able to monitor and match up with their activity, then, you are better off being on LinkedIn.

Budget

Advertising on LinkedIn costs several times more than advertising on other social media networks. This is primarily because of a small number of highly targeted audiences. This raises competition by B2B advertisers for the opportunity to reach out to this niche.

According to AdStage, the following were LinkedIn average figures for advertising in 2017:

- Average CPM for LinkedIn Ads was $8.39
- Average CPC for LinkedIn Ads was $6.50
- Average CTR for the LinkedIn Ad was 0.13%

You must factor these Ad rates into your budget. A small B2B company should be more conscious of how these rates could deeply impact on its marketing budget and the likelihood of it not being a better way to optimize its marketing ROI.

Who should advertise on LinkedIn

Generally, the following are the most potent areas to be advertised or considered on LinkedIn:

- High-value B2B products and services
- Recruiting efforts
- Higher Education

High-value B2B products and services

In case you are offering high-value B2B products and services such as subscription services e.g. SaaS, then you are likely going to hit big on your LinkedIn advertising.

Recruiting efforts

If you are HR Consultant or company seeking high-level professionals, LinkedIn is the place to have the most impactful advert. With advanced filtering mechanism, such as education level, professional level, years of experience, and salary level, job title, age, among others, you can easily target specific potentials and are likely going to get the right kind of staff.

Higher education

If you are an institution of higher learning and offer an advanced degree program; for example, masters, doctorate, or executive degree programs, then targeting those with at least a bachelor's degree will be the most appropriate thing to do.

On the other hand, you could be a large university with huge alumni. You want to connect with your alumni for particular academic and social programs such as sponsorships, then targeting those who attended your particular institution of higher education would yield optimal impact.

Who should not be advertising on LinkedIn

LinkedIn is not suitable for all kinds of advertising. Thus, there are those advertisers who may not get adequate ROI for their advertising campaign due to the unique nature of the LinkedIn audience. The following are some of the entities that should weigh their position well before considering advertising on LinkedIn:

- B2C companies – These are better served by other social media networks such as Facebook and Twitter. This is primarily because LinkedIn does not have such targeting parameters that you can use to target your audience. Furthermore, the cost of clicks on LinkedIn is several times higher than that of Facebook and Twitter, which makes it poor for ROI on small-scale sales.

- Smaller deal sizes – As a rule of thumb, for a deal that is lower than $10,000, it is more costly to use LinkedIn as you may spend more on ads than your gains.

- Broad targets – LinkedIn is for a specific target. If you are targeting broad base (such as gender, or age), then it is better to use other social media channels.

- E-commerce – Ecommerce is ideal for 'buy-now' CTAs. However, most of LinkedIn audiences are not inclined to such as they attend the platform to network.

- Commodities – Commodities generally fetch lower CTR (Click-Through Rate) on LinkedIn. It may seem appealing to offer high-end commodities such as bank loans, real estate, and insurance policy, among others. However, most of the professionals on LinkedIn are already networked with contacts offering such services. However, there is a chance to harness young professionals who are just entering the trade.

Advantages of LinkedIn Advertising

Advertising on LinkedIn has certain unique benefits. The following are some of the benefits you gain by advertising on LinkedIn:

- A broad variety of ad types – As we have seen before, LinkedIn provides a broad range of unique Ad types. This helps you to package your Ad in the most effective manner.

- Targeting precision – Due to its advanced filtering options, you can have more specific targeting which yields higher ROI
- Specialized analytics tools – With advanced tracking and analytics tools, you can easily measure the performance of your Ad campaign in terms of lead generation, conversion rate, and ROI
- Flexible budget – With LinkedIn, there is neither minimum nor maximum cost when it comes to advertising. You can create your own customized budget and finance your campaign as much as you can.

USING LINKEDIN GROUPS

LinkedIn groups are the most powerful of any group on social media network in terms of professional competence. LinkedIn being a professional networking site, it simply means that members on LinkedIn are likely going to connect with those whom they share some professional value – be it belonging to a similar profession or having similar or mutually beneficial professional interests.

This means that to be able to touch base with your target audience, you have to belong to relevant groups.

Why you should be using LinkedIn groups

There are five core reasons that would drive you to use LinkedIn groups for your marketing effort:

1. Enhance your credibility – Showing mastery of your core area of competence is important in thought-leadership. This helps to strengthen your Goodwill and thus generate a positive response to your marketing campaign. Sharing relevant content, providing incisive ideas and commenting on latest industry development can help to boost your thought-leadership.

2. Expand the reach of your content – Content marketing is important to your marketing endeavor. LinkedIn groups provide you with an avenue where you can share your content.

3. Generate ideas for new content – With LinkedIn groups, you can learn what attracts attention by learning about the most-talked-about topics. This can help you fine-tune your content strategy.

4. Learn about your target audience – The easiest way to learn your target audience is by being closer to them as much as you can. LinkedIn groups provide such high level of proximity. With proper listening, you can easily tell your audience's desires, needs, challenges, fears, and pain points. This can help you in your entire marketing endeavor right from product design, branding, and delivery.

5. Find the crowd for your next event – You will occasionally require hosting seminars, PR events, and networking events. You would definitely want to reach out to those whom your events can have a significant impact. Inviting members of your group will not only help you gain the required crowd but also enhance your relationship by building stronger bonds.

The advantages of LinkedIn groups to your marketing endeavor

LinkedIn groups can serve your marketing endeavor in various ways. The following are some of the advantages that you gain from LinkedIn groups:

- Groups let you connect to a thriving community

LinkedIn groups provide some of the best professional communities that you can have on social media. Members are highly committed. This saves you the effort of creating your own community from which to draw potential leads.

- Groups help in establishing your network

LinkedIn groups help you establish personal connections and networking. You can leverage on this network to boost your marketing effort.

- They can be used for lead generation

Leads come from the community. Groups help to bring those leads closer to you. Furthermore, you can still draw more leads outside the group through members' connections.

- Groups can drive traffic to your site

Once you have established a reputable relationship with members of your group, it is much easier to request them to visit your site thus increasing your site's traffic and potential leads.

- Groups can be used to establish yourself as a thought leader

Within groups, you have the power to generate and foster ideas. This helps to influence people's mindset. When people find your thoughts compelling, they become aligned with your thoughts and as such, you become their thought leader. You can use this thought leadership to persuade them to throw their weight behind your brand.

How to find and join the right group

There are thousands of groups on LinkedIn. It can be a daunting task to find the one which fits your needs. The following steps can help you identify the right group to belong:

- Establish your goal – Be mission-specific when it comes to finding the right group. Having specific goals will help you narrow down your search criteria to only those groups that fit your goal. For example, do you want to find peers, potential customers, or potential vendors?

- Use the right search terms – With your goals in mind, you can define terms to use which will enable the search engine to come with groups that match them.

- Take a lead from those you follow – Find out where the bulk of your followers come from. Most likely, the groups that they come from will have something in common with your needs. Evaluate them and see the ones that you can join

- Read group descriptions – A group's description can enable you to establish what the group is all about. Some groups have specific kind of people that they want to join them. This will help you determine whether the group is right for you or not.

- Gauge the quality of group discussions – Within the shortlist of your potential groups, follow up on their discussions and gauge the quality in terms of relevance, depth level, and engagement

- Join a few of your selected groups – Sometimes the best experience is insider experience. Being in groups will help you to understand their inner workings and your comfort level. If you do not feel comfortable being within a certain group, you can leave it.

- Join groups where your customers and peers are more likely to be – If you want to boost your marketing effort, you would want to associate with groups where your customers are.

- Expand your connections – it is much easier to join a group and receive attention than to be followed by members of that group. Thus, you might need to use LinkedIn Connections Service to grow your following.

How to create a successful LinkedIn Group

If you find out that none of the groups caters for your specific interest, you are free to create your own group. You could find people who have similar interest and need for such a group.

To be able to create a successful LinkedIn group:

- Pick a topic that resonates with your customers' needs

Your group should focus on a topic that has a natural connection to your brand but not actively promoting your brand or company. For example, if your brand is a fitness gear brand, your topic should be about fitness workout in general rather than your specific brand. This way, your audience will begin to gradually and naturally start connecting with the topic and your brand. This has a more valuable long-term impact than a sales pitch.

- Create your LinkedIn Group

With the right topic in mind, you can now go ahead and create your LinkedIn group. Simply navigate to your LinkedIn Groups and click on "Create Group". You will be directed to a form to fill in the group details which include Group title, Group logo, description, group rules, and group membership type (standard, or unlisted).

- Set up message templates

Message templates help to create custom messages that would automatically be sent to people interested in joining your LinkedIn Group.

- Invite your connections and grow your group

Once you have set up your group, you can now invite your connections to join the group. Also, encourage group members to invite their connections into the group.

- Start discussions and be active

Make posts, the first being the Welcome Post where you welcome new members to the group, tell them what it is all about and remind them to observe the rules for the good conduct in the group. Start and engage in discussions based on the question-and-answer format of posts. You need to be actively engaging them and post more since your posts and discussions will set the benchmarks of what you expect from group members.

- Moderate all posts and remove spam

Moderation helps to ensure that there are no spam posts. Spam posts are a turn-off and will lead to members quitting the group. Set LinkedIn auto-moderation system to flag promotional content.

- Take advantage of the promotional area

Linked in has a "Promotions" section. This is where you share articles and other kind of content that link back to your site. It is not a place for a sales pitch.

- Build connections and Follow Up

Keep yourself active in your group through posts, engagement, and moderation. Be respectful of the opinion of other members even if you tend to disagree with them.

How to optimize your opportunities within your group

The following are ways by which you can optimize your opportunities within your group:

- Follow the rules – Observe the group's rules. Learn the group's etiquette and keep observing it.

- Become engaged – Respond to other people's posts and comments. Let it not be all about you.

- Post as an individual, not a company – people like engaging with other people, not entities. People do business with people. This should inform the way you post.

- Stay relevant – Provide content that benefits the group as a whole, not just you. Let the content create value for the group and other members. Self-centeredness turns-off others.

- Avoid regurgitating same content from one group to the other. Some members of your group are also members of some other groups that you belong. If they encounter the same post again in another group, they will be pissed off and this lowers your reputation.

- Be an active, contributing member – Dormancy works against your marketing endeavor.
- Showcase your expertise – Share information, content and ideas about your core area of specialty. Establish yourself as both a thought leader and an expert in your field.
- Take polls and ask questions – Posting a poll and asking questions is the best way to gauge your audience opinion. This not only boosts engagement but also helps you have the right perspective on them and device appropriate marketing strategy.

HOW TO USE LINKEDIN ADS

Ads are a very important marketing tool that no firm can afford to ignore. LinkedIn has plenty of unique Ad types that you can use to advance your marketing effort.

To be able to use LinkedIn Ads effectively, the following are the important steps you need to follow:

- Establish your buyer persona
- Determine the right approach to target your buyer persona
- Decide the right Ad(s) to target your buyer persona
- Select your bidding option
- Establish your ad budget
- Launch your Ad

Establish your buyer persona

At the end of it, you want your Ads to convert. This cannot happen if you are not targeting potential buyers. Thus, you need to establish your buyer persona so that you can have targeted Ads. The following are ways to establish your buyer persona.

- List down all traits and demographics of your ideal buyer e.g. age bracket, education level, urbanity, income level, among others
- Speculate about their likely pain points, motivations, and concerns
- You can also use Facebook Audience Insights and Google Trends to determine additional info about your buyer persona

Determine the right approach to target your buyer persona

Once you have established your buyer persona, the next thing to do is to find the right packaging of your Ads in such a manner that will inspire their positive response to your CTA (Call-to-Action)

Decide the right ad(s) to target your buyer persona

Depending on your approach, the following are the types of LinkedIn Ads that you can use:

- LinkedIn Self-Serve Ads
- Sponsored content
- Text Ads
- Sponsored InMail
- LinkedIn Display Ads

- LinkedIn Dynamic Ads
- LinkedIn Marketing Partner Ads
- LinkedIn Video Ads

LinkedIn self-service Ads

LinkedIn allows you to publish your own Ads through its Campaign Manager. You can schedule the campaigns in advance, target your desired audience and be able to get metrics of those who click your Ads including their industry, job function and level of seniority.

Sponsored Content

Sponsored Content is a facility that lets you share posts on your Company Page with a targeted audience on LinkedIn. This is a great feature as it allows you an opportunity to customize your Ads within your content and put it in your most desired way.

Sponsored content helps you to:

- Attract more followers to your Company page
- Extend your content's reach
- Integrate LinkedIn's lead generation forms
- Capture views and clicks on mobile, tablet and desktop

Text Ads

Text Ads are a common form of advertisement online and not necessarily unique to LinkedIn. However, on LinkedIn, they spruced up to include more features such as a desktop-only option feature that lets you have a compelling headline, concise description, and customized eye-catching image.

Text Ads allows you the following benefits:

- A quick start
- Budget-friendly customizations
- Conversion tracking
- Focused target audience based on precision filters

Sponsored InMail

This feature is natively unique to LinkedIn that allows you to deliver personalized messages to a targeted LinkedIn audience through LinkedIn Messenger. It is designed for maximum attention such that the message alerts come when the targets are actively online.

Sponsored InMail can be used to:

- Make personalized invites to events and webinars
- Promote downloadable content such as whitepapers and eBooks

LinkedIn Display Ads

Display Ads are a common mode of online advertisement. LinkedIn has customized Display Ads, which you can buy from your preferred advertising platform, or auction (both private and public). They are highly versatile in terms of multimedia capabilities as you can incorporate text, images, audio, and video for maximum attraction and experience.

You can use Display Ads to:

- To quickly reach out to a wider audience during the initial launch of your campaign
- Strategically locate and expose your campaign to high traffic sources on LinkedIn
- Cut across diverse categories of an audience including thought-leaders, opinion-shapers, influencers, and decision-makers

LinkedIn Dynamic Ads

Dynamic Ads are highly intelligent ads that provide customized Ad message based on the target's behavior. For example, if a prospect is searching for a job vacancy within your industry, the dynamic Ad will customize a message for the prospect regarding a job at your place. Dynamic Ads are highly personalized and customizable.

You can use Dynamic Ads to:

- Reach out to the most influential members of your audience
- Build and cultivate a strong relationship with the targeted audience

- Deliver highly personalized and compelling message to boost the desired action
- Attract specific action through customized CTA
- Widen personalized reach through a wide range of criteria
- Grow a loyal following to your Company's page.

LinkedIn Marketing Partner Ads

These are Ads created by LinkedIn marketing partners. If you feel that you are not confident, or you do not have an in-house team of expert marketers, then you may seek LinkedIn marketing partners to carry out Ads on your behalf. They do have the right experience to deliver Ads as per your expectations.

Video Ads (coming soon)

LinkedIn has been experimenting with video Ads, which are currently in beta release. They are projected for a full launch by the end of the first half of 2018. Like other native LinkedIn Ads, they will be available through LinkedIn's Campaign Manager.

Choose Your Bidding Options

Once you have selected your preferred Ad type, then next step is to choose your bidding option. There are two traditional types of bids:

- CPM (Cost Per Impressions) – This is costing based on impressions (views, exposure) on your Ad. One unit is worth 1000 impressions. Thus, every time the impressions reach 1000, you are billed a certain agreed amount. CPM is ideal when you are targeting brand awareness. Here clicking is not as important as reading the message.

- CPC (Cost per Click) – This is costing based on the number of clicks made on your Ad. CPM is ideal when you are targeting conversion and thus people have to click to make a buy decision.

Establish your ad budget

Budgeting is crucial for your marketing campaign. If you are starting out, it is better to set a daily budget so that you can monitor the performance of your various Ads before deciding on the type of Ads that you are going to promote on a long-term basis. It is only after being assured of the type of Ad that does well that you can set a long-term budget on it. Never assume that the type of Ad that appeals to you most will also be the most appealing to your audience base. Carrying out A/B Testing with a pair of Ads will help you save your budget from being spent on the wrong type of Ad.

Launch and run your ad campaign

You can now go ahead, launch your Ad, and run your Ad campaign. To make a successful launch and Ad campaign, you will need to consider the following steps:

- Go to <u>Ads</u> platform and select "Create Ad" to launch your Ad. You will be directed to your Member dashboard (Campaign Manager) where you will be required to enter your billing information.
- Click on the call-to-action (CTA) button to create a campaign. Select the kind of Ad you want to create.
- Set your Ad's basic parameters These include:
  - Ad's primary language
  - Campaign name
  - Targeted demographic
- Establish your Ad's media and format - This is now about building your actual Ad based on the parameters. Here you will consider factors such as format, layout, among others. In building the Ad, the following will be required:
  - Ad headline – This is a maximum of 25 characters
  - Ad body – This is a maximum of 75 characters. It should be targeted to a specific buyer persona. Its content should be relevant to both the buyer persona and the landing page you are directing them to.
  - CTA – Have actionable CTA to boost click-through
  - Value – Make a value proposition. The value should be in terms of what the audience benefits by responding to

CTA. This could be a special discount, etc. It should also have time value in the sense that it draws urgency to take action.

- Testing – Carry out A/B Testing by using multiple variations of your Ad in each campaign so that you can tell which of them has a higher impact on future campaigns.
- Target your Ad – Use LinkedIn filtering parameters to target your Ad for the 'right clicks' and maximum impact. This is optional. Specify parameters relevant to your Ad campaign. In this targeting, you can specify:
  - Location
  - Company
  - Job title
  - Member schools
  - Member skills
  - Member groups
  - Gender and age

MONETIZING YOUR LINKEDIN FOLLOWING

Monetization is the reward of every marketing endeavor. You obviously would like to be rewarded for your effort. LinkedIn offers unique monetization opportunities that can be of great benefit to your business. However, being a professional network, it means you have to employ uniquely different marketing and monetization strategy than you would on other social media networks.

Understanding LinkedIn Monetization Strategies

Before you engage in monetization effort, you first need to understand the best approach to LinkedIn monetization. Primarily, LinkedIn monetization is less about content monetization and more about monetization of smart networking. This is radically fundamental.

Content monetization simply means using content as a sales vehicle. It primarily focuses on affiliate marketing, information products, advertising and such like. This is less of a focus when it comes to professional networking where ideas and influence through thought leadership are the primary focus. Nonetheless, this does not necessarily mean that you cannot use content monetization strategies. It simply means that they should be secondary in nature. LinkedIn has a publication feature that can allow you to utilize content monetization.

Understanding LinkedIn Channels

LinkedIn has features that allow you to connect with others, form core interest groups, promote yourself, publish content, and promote products, among others.

The following are the main LinkedIn channels that you can use to advance your monetization endeavors:

- LinkedIn Profiles – LinkedIn profiles are places to showcase yourself, your company, your services or products. You can link each to your personal blog, company website, landing page, respectively.

- Long-form publishing – LinkedIn allows long-form publishing. You can use long-form publishing to provide infographics, thought-provoking multimedia content, among others.

- Groups – Groups provide an excellent opportunity to categorize your audience according to specific interests and criteria. They are a great place for closer and deeper engagement.

- Advertising – LinkedIn, unlike other social media, provides you an opportunity to target your advert to anyone within the network based on a number of filtering options.

How to use your LinkedIn account to monetize your following

LinkedIn is a great place for established brands to make money. There are two phases involved in your monetization endeavor:

- Preparatory phase, and
- Monetization phase

Preparatory phase

In this phase, you are laying a solid groundwork for monetization. You are building the monetization factory. How far you will succeed in the monetization phase depends on how successful you were in the preparatory phase. This is why it is important to spend considerable time in the preparatory phase until you are confident enough that you are ready for take-off to the next phase. Unlike other social media, LinkedIn is not a place for 'hit-and-run' kind of affair. It is not a place for sprinters but marathoners. You have to focus on the long-term.

In this, you need to:

- Build your LinkedIn profile – It is common to have a basic profile like so many that exists on LinkedIn. However, when it comes to monetization, there is more work to be done. First, your profile must be complete. Second, it must be aesthetically appealing (good-looking photo). Third, it must prove compelling value – in terms of great resume (what you have done) evidenced by portfolio (what you have achieved/ your success story).

- Tell a brand story – In this regard, you need to articulate clearly your vision, mission, and the journey so far traveled in terms of accomplished milestones and where you are heading. It should be inspiring enough for your audience to want to be part of your journey.

- Build bonds – This is the core essence of LinkedIn – to create professional networks. However, being connected is one thing and building bonds is another thing. The connection is the first step, which most people on LinkedIn achieve. Building bond is the ultimate step that most do not bother to endeavor. Building bond is about creating that strong affinity, attraction, and ties through a common fabric. You must create something, which people within your network feel connected to, and find it hard to do without it. This could be a great idea, a great value, or a great service.

- Grow your mailing list – Use your connections to grow your mailing list. It is from this mailing list that you can leverage your monetization effort. It is extremely important to growing your mailing list before launching your monetization campaign. This allows your connections to get accustomed to your mailing list for valued services rather than a sales pitch. When you later on introduce measured sales pitch, they will not get offended.

Monetization phase

This is the reaping phase. It is a phase where you direct your focus on making monetary gains from your effort at the preparatory phase.

- **Sell Info Products**

You can leverage your LinkedIn following to sell info products. These info products could be white papers, eBooks, reports, webinars, and free training. The following are some of the ways by which you can achieve this:

➢ Create free content as an invitation to the wider end of your funnel – you can create white papers, eBooks, and reports and free training as a way of subtly proving your product's concept to your audience. Share links to your squeeze page with your connections and groups.

➢ Connect with influencers in your niche and inspire them to share out your info products.

- ➢ Offer discounts to LinkedIn members on your info products – Create special offer code for your new info products and share with your connections and groups
- ➢ Invite people to your webinar – Provide free webinars that offer high-value education. You can have a sales pitch within and at the end. People will tend to respond positively as a psychological reward for your great service.
- ➢ Take advantage of the Projects Section of your LinkedIn profile to add your new product – This is the only section of your profile that has the room for you to link directly to your website
- ➢ Seek reviews of your info products from influential members of your group or connections

- **Sell Physical or Digital Products**

You can tactically sell physical or digital products on LinkedIn. The following are some of the ways by which you can do this:

- ➢ Exploit the Products Section of your LinkedIn company page to create product listings and clickable banners linked to your landing pages, product demo videos, and 'learn more' pages.
- ➢ Take advantage of your LinkedIn summary section to put CTA presentations and commercials.

- Create or get involved in groups whose members are more inclined to buy your products. Present your product as a solution to their unique needs where you find them seeking a solution, which your product can adequately serve.
- Inspire people to click on the 'Recommend' button on your product page. This helps to boost your social proof on LinkedIn

- **Increase Book Sales**

LinkedIn, being a professional network where ideas and thoughts are highly encouraged, it follows that books (being nothing but a package of thoughts and ideas) would also be welcomed. However, it is important to ensure that your book is highly targeted to its likely beneficiaries.

To increase your book sales:

- Inform your connections the launching date of your new book. If possible, invite them to the launching event, either physically or through webinars. You can send this information through customized private messages.
- Incorporate your books to the 'Publications" section of your LinkedIn profile. You can link them to your Amazon page or online store for them to click and buy.

- ➤ Join groups whose core interests are relevant to the topic of your book. This allows you to discuss the topic and associate your book with the topic. It is an easier way of informing them about your book without appearing to augment a sales pitch. Listen to cues that inform you that the moment is opportune for you to mention your book.

- ➤ Create a video summary or trailer of your book. Launch it on your YouTube channel and then insert it as a Media element in the summary section of your LinkedIn profile.

- ➤ Request an influential member of your connections to review your book and share it with his/her audience.

- ➤ Write a post about your book on your blog and add a LinkedIn share button to the blog post. Click the LinkedIn button to share it with your connections.

- ➤ Take advantage of targeted LinkedIn advertisement to reach out the book to a wider audience than your connections. Of course, you will have to use filters to target the right audience.

- **Find Direct Advertisers and Sponsors**

In case you are a web publisher or blogger, you may need direct advertisers and sponsors to buy space on your website or blog. LinkedIn is a good place to find those in need of such services. You can find such opportunities by:

- Taking advantage of LinkedIn's search page to find potential sponsors and advertisers
- Follow-up on advertisers on the LinkedIn footer, sidebar and other areas on the platform. They could be in need of such other places on your website to advertise their products and services
- Reach out to potential companies and make a cold attempt by reaching out their employees on LinkedIn. You can warm up and nurture the connections just as you would nurture a lead.

- **Promote Affiliate Products**

If you are an affiliate marketer, LinkedIn is a great opportunity to promote your affiliate products. However, it is important to note that you will have to be more creative than on other platforms as most members tend to perceive affiliate links negatively if not from people they trust and have deeper engagement with. Thus, you have to do more to build a relationship first and leverage that relationship for affiliate marketing.

The following are ways by which you can promote your affiliate products:

- Use sponsored long-form posts to insert your affiliate products links

- ➤ Take advantage of the Products Section of your Page to list your affiliate products with relevant affiliate links
- ➤ Use LinkedIn's Ad Facility to advertise your affiliate product
- ➤ Create your affiliate product's review posts and place them in the Publications section of your LinkedIn profile.
- ➤ Request your influential connections to write a review of your affiliate product and let them share it with their connections. Also, do share the same with your groups.
- ➤ You can take advantage of your group by sending them emails about latest product reviews, new product launches, discount codes, limited time offers, among other promotional content.
- • Use LinkedIn Advertising.

Advertising is an opportunity for anyone to reach out to a wider audience. You can advertise any of your products and services to reach out your target audience on LinkedIn. However, create one campaign for each of your product or service in order to have a more accurate target.

- • **Offer Consultancy Services**

LinkedIn is a great place for freelancers and consultants to offer their professional services. The following are some of the ways by which you achieve this:

- Add relevant skills to your LinkedIn profile. This will introduce you to your potential clients by letting them know that you are qualified to offer certain services.
- Utilize the Service Section of your LinkedIn Company page by posting a listing of your services on offer. You can optimize each of the listings by adding banner images with links to your website, video testimonials, and referees.
- Use SEO keywords relevant to your skills to optimize your LinkedIn profile. This will allow discoverability of your skills by those looking for them.
- Use LinkedIn's advanced filtering tools to find groups where your potential clients are likely to be and join in. This way, you can participate in the group discussions, build rapport and introduce your services.
- Watch out for leading questions indicating that someone is in need of your services and grab the chance to lead the person to your services.
- Link-up with those who offer complimentary services and join their groups. For example, if you are an editor, you may need to link up with freelance writers; if you offer plumbing services, you may link up with Home and Office building contractors; if you are an SEO expert, link up with web developers.

- **Get Hired**

Finding jobs and being hired is the traditional function of LinkedIn that continues to remain relevant over time. It is a place where jobseekers expose themselves to potential recruiters. The following are the ways by which you can boost your chances of being hired on LinkedIn:

➤ Complete your personal profile and make it up to date with your latest skills and achievements.

➤ Take advantage of the Summary section of your LinkedIn profile to express your passion, showcase your accomplishments, and prove what you are ready to deliver to your potential employer.

➤ If you are targeting a company for a job, follow the company's page and deeply engage with it by commenting on their posts. Also, join groups in which the company's senior-most staffs relevant to your field are. Keep checking their Career's tab and the Job section of the company's page for likely openings, which you can respond to them.

➤ Premium account for job seekers is a great feature that you should exploit, as it will push your applications on top list to make it stand out. The premium icon on it will also help to accentuate its case.

GROWING YOUR LINKEDIN FOLLOWING

A large LinkedIn following simply means a wider opportunity for exposure and a greater lead potential. Thus, it is important for you to grow your LinkedIn following to optimize your marketing effort.

The following are some of the ways to grow your LinkedIn following:

1. Leverage your LinkedIn profile and company page

Your personal profile and company page are great tools to increase your LinkedIn following. You can optimize these by:

➢ Making sure that your LinkedIn profile is complete and up-to-date with the latest information.

➢ Optimizing the first 150 characters of your page description for SEO and click-through. It is the first 156 characters that will be displayed on the SERP.

➢ Having a catchy image on your company's home page to attract visitors' attention.

- Having a QR code on your business card to direct people to your page. Also, include your page's link on your business card.

- Posting your page updates during peak hours such as Morning or Midday on Monday to Friday. This will increase impact on viewership.

- Having multiple showcase pages on your profile each pointing to a particular dimension of your company including departments, product categories, and corporate social responsibility, among others.

- Using your company page as your current place of work. This will automatically cause your logo to be displayed on your personal page, which increases the chances of a click-through.

- Using LinkedIn Connect to synchronize all your contacts on LinkedIn. Request new connects to follow your page.

- Assigning several trusted administrators to your page. Encourage the administrators to post engaging content on your page and share updates with their connections.

- Naming your page using your business full name. This is a way to market your business and create brand awareness. More people who know your business but do not know your company page or personal profile will easily discover you and connect. This will also enhance your ranking on Google search engine.

- ➢ Posting your job vacancies on your page. This will increase your page following as potential job seekers will be on the lookout
- ➢ Letting your LinkedIn Page link be part of your group signature rather than your web page. This is much easier for your connections to reach without having to leave LinkedIn.

2. Leverage your company employees

- ➢ Encourage your employees to join LinkedIn and add your page as their workplace. This automatically makes them your followers and encourages their followers to join you.
- ➢ Encourage your employees to actively and positively engage with your page's content. This is a good social proof of your page to other followers.

3. Leverage your content

- ➢ Make sure that your page is not a skeleton before requesting for followers. Let them find enough content to keep them occupied, engaged and knowledgeable of what your page is all about.
- ➢ Take advantage of LinkedIn Pulse to encourage comments and shares on each post.

- ➢ Aim at least five posts per week to keep your presence felt and your followers engaged. The more frequent you post (not compromising on the quality or overwhelming your readers) the more reach you attain.

- ➢ Optimize your content through lists to amplify reactions. Lists, especially "the best-of" lists help to boost amplification by close to 40% (according to LinkedIn).

- ➢ Take advantage of the images to attract attention. Make sure that your post has at least one relevant, high quality and attractive image.

- ➢ Use your page to share inspiring company news. Most professionals are on LinkedIn to eavesdrop on latest company news.

- ➢ Provide industry-related insights to attract more following. This is what most professionals are on LinkedIn for.

- ➢ Share useful content with your followers. They will greatly appreciate this extra service. Furthermore, you will be able to gauge their interest using such shared content and thus calibrate yours to the same wavelength.

- ➢ Be an early listener to the trending topics so that you become among the first to offer breaking insights.

- Provide credible and valuable links within your content. When you do this persistently, your followers will grow a habit of looking up for the links to click and share with their connections. This will help you to earn new connections from their sharing.
- Market your LinkedIn platform. You can do this by creating a blog post about some unique feature or unique connections that need to be followed on LinkedIn (e.g. 10 Professional Financial Advisors on LinkedIn". This will not only increase traffic to your blog, excite more connections, but also improve your page rank on search engines as the search engines may pick a part of your blog post as a LinkedIn keyword (e.g. "Financial Advisors on LinkedIn".
- Do not forget to include CTA at the end of your blog post requesting readers to follow you on LinkedIn.

4. Leverage your other social media networks and online assets

- Add LinkedIn button to all your online assets including blogs and web pages.
- Add your company's LinkedIn Follow button to your website and blog

- Use your other online marketing channels such as social media channels (Facebook, Twitter, YouTube, etc) chatting platforms (such Whatsapp, Skype, etc) and emails to increase your LinkedIn page visibility.
- Insert a link to your LinkedIn page in your email signature
- Add your company's LinkedIn widget to your website's sidebar. This will stream relevant information about your company and thus encourage click-through. It also has a Follow button to encourage following of your company's LinkedIn page.
- Share videos on your YouTube channel with your LinkedIn followers. This boosts engagement and shares which attracts a new following. According to LinkedIn, videos create twice as many reactions (comments, shares, etc) compared to posts without videos.
- Guest blog and make sure that you insert a link to your LinkedIn company page in your bio on the guest blog post you make.
- Take advantage of SEO to optimize your LinkedIn pages for searchability. Make sure you include relevant keywords in your long-form posts and your page description.

5. **Leverage your engagements**

➤ Have active engagement in LinkedIn groups. Share your thoughts and insights regarding your industry and profession. Build thought-leadership around this core area. You will gain more trust and following as an industry expert with a lot to offer.

➤ Make your current connections excited before attempting to grow more. Otherwise, you may achieve short-term success and long-term loss as more connections exit.

6. Leverage the LinkedIn tools and features

➤ Keep attention to your LinkedIn's content marketing score. This score will inform you of what percentage of your following is being reached by your content marketing campaign. The score also has metrics that enable you to know how you are performing against competition within your industry. This grants you an opportunity to strategize and improve your content marketing campaign effort.

➤ Keep your eyes on LinkedIn analytics and performance metrics with regard to how your posts are doing in terms of inspiring your audience. You can then emulate your best performing posts as your standard. This is also a good way to gauge your followers' core interests and match strategy towards them.

7. Leverage your LinkedIn group and existing following

➢ Inspire your connections to follow your page and share it out with their own connections

➢ Make more connections by following more pages. The more expansive your network becomes, the greater the potential of growing your page's following.

➢ Identify your complementary businesses in your industry and follow them. Engage with them and share their content. They are more likely to return the favor and join you in wider partnership as you have mutual interest.

➢ Identify those members that appear to be good at consuming content and connect with them. They are more likely to consume your long-form content and establish deeper connections with you. They are also more likely to summarize key points on your behalf as they share it out with their connections. Encourage those who show positive response to do so.

➢ Take advantage of questions to raise your audience interest and gauge their sentiments. According to LinkedIn, you are likely going to get 50% more comments on posts with questions.

- Create or Join an industry-related LinkedIn group. Share relevant content and valuable information as a means of introducing an engaging discussion.
- Keep self-promoting content to a minimum. Employ Pareto's 80/20 rule such that 80% of your content should be non-promotional content while only 20% should be promotional.
- Segment your followers based on their key interests so that you can have more targeted posts that raise engagement. This will encourage a higher level of engagement, more shares with their connections sharing similar interests and the net result will be more following.

AUTOMATING LINKEDIN INCOME

Automation is the best way to scale up your LinkedIn monetization effort. Without automation, you will be dwarfed by the overwhelming amount of work that you need to do just to maintain your current status, leave alone growing.

The advantage of automating your LinkedIn Income:

- Cost cutting – With LinkedIn income automation, you are able to cut costs in terms of time and money. In some core functions, using automation tools can achieve you much more than what you would achieve from hiring 20 people.

- Structured approach – Humans tend to be erratic and unstructured depending on many biological factors. However, automation tools will be more structured as they are not susceptible to human biases. This way, you can be able to get a more precise outcome.

- Larger-than-life presence – Automating certain LinkedIn functions such as visiting your followers' profiles, making personalized emails can make you appear larger-than-life.

Automation tools can work 24/7, unlike a human being. The can also visit multiple profiles at the same time.

Popular LinkedIn Automation Tools

- LinkedIn Sales Navigator – This is a relationship-building tool that helps you identify potential leads through its recommendation mechanism and helps you to draw your leads ever closer to the sales funnel.
- Rapportive – It gathers data from LinkedIn profiles social media network accounts and web URLs to provide you crucial information about leads' contacts and other important personal details.
- LinkedIn Plugins – These are plugins that help to enhance your LinkedIn account functionality. Some of the popular LinkedIn plugins include;
  - ➢ Company Profile
  - ➢ Follow Company
  - ➢ Company Insider
  - ➢ Alumni Tool
  - ➢ Member Profile
  - ➢ LinkedIn AutoFill
  - ➢ Share

- LinkedIn Small Business Center – This is a LinkedIn in-built facility dedicated to helping small businesses grow. It helps you to establish a brand presence, connect with the target audience and use content marketing to engage with them.

- Crystal – Crystal is a profiling tool that reviews LinkedIn profile and provides insight into the person's personality. This helps you decide on the best way to approach a given prospect. As such, it is a buyer persona's automation tool. When you want to draft email and other communication, Crystal will recommend to you the best approach based on the identified persona.

- LeadFuze – This is a leads prospecting tool. It helps you generate a list of potential leads. You can use its lead builder to send personalized emails and follow-ups automatically.

- SalesLoft Prospector - This tool enables you to personalize your communication based on email tracking, use built-in sales dialer to increase sales performance, and gather incisive analytics.

- Outro – It uses "relationship strength algorithm" to help you establish qualified leads.

- Salestools.io - This download your LinkedIn lists to Excel. With the Excel, you can easily manipulate the data to suit your particular need such as ranking prospects in order of importance. You can also use the tool to extend personal outreach messages and track your leads activities.

- eLink Pro – This tool automates profile visitation. Most of the time, people visit your profile and connect with you based on your visit to their own profiles. The more profiles you visit the higher the likelihood of being visited back. This tool does the visitation work for you.

- Discoverly – This tool provides you with information about your leads' other social media activities so that you are able to know them better.

- Guru – This tool helps to build a profile of potential leads by finding and availing data such as prospect's competitors, prospect's current customers that work in the same industry, sales data related to the prospect's industry, among others.

- FollowingLike – This tool helps you manage not only LinkedIn but also multiple social media accounts.

- eGrabber – This tool helps to find missing information about your potential prospects such as emails and phone numbers of your LinkedIn connections. It does this by extracting this data from various other sources online including other networking sites and online directories.

- ProTop – This tool helps you to search the right buyer persona and generate leads. It automatically visits profiles of LinkedIn users that match your set criteria and helps to drive attention and traffic to your profile.

- InBoardPro- This tool helps you to manage multiple LinkedIn accounts, generate leads, and get right candidates for recruitment.

- Scout – This marketing automation tool is used to carry out sales prospecting. With it, you are able to get personal contact information of your top LinkedIn leads.

- Nimble – This is a CRM tool. It uses Relationship Intelligence to enable you to make real-time discovery of detailed profiles and automate social profile matching.

- LinkedIn Dominator – Helps you to add connections based on keywords.

Outsourcing as part of the automation process

Automation tools cannot handle all the human functions. Roles such as responding to posts and comments require a human touch. Engagement is one key area that you will need not to surrender to automation tools as you are likely going to get negative results. However, due to automation of other functions, the volume of engagement requirements flares up to such a level that you cannot manage alone. You will need to hire someone to help you carry out the engagement function. Luckily, many trained and experienced virtual assistants exist who can do this work on your behalf. Outsourcing some of the engagement work to virtual assistants would help you complement what automation tools are doing.

You can easily get freelance virtual assistants from Upwork, Freelancer, People Per Hour and Fiverr, among other freelancing sites.

CONCLUSION

Thank you for acquiring this guide and reading it to the end.

It is my sincere hope that information provided in this guide has enabled you to use LinkedIn B2B marketing to generate qualified prospects and obtain clients. It is also my sincere hope that you have been inspired enough to recommend this guide to your friends and colleagues so that they too can obtain a copy of it and that it can help them succeed in their LinkedIn income generation endeavor.

Again, thank you for acquiring this guide.

Have good luck!

Email Marketing

A Step-by-Step System to Build
Passive Income Using Email
Marketing

DISCLAIMER

What is Email Marketing?

In one line, email marketing is how you send advertising and promotional messages through emails.

Emails provide direct access to a vast market. Almost every person uses an email account. Hence, you can use this platform for cheap, fast and effective marketing. It allows you to connect businesses, products, and services to relevant consumers. And earn in the process too!

Email is known among marketers as a highly flexible tool. You can create simple or flashy messages and send it to an extensive network with a single click. Emails can contain multimedia, texts, videos, links, images and many other features. The message generation depends on marketing goals. Some messages require simple texts, while others need links, images, and other flashy elements.

Targeted marketing is the most significant benefit of email marketing. Other tools such as television, radio, or even print don't provide a precise demographic to target. However, this is not same as in email marketing. Personalized messages are created to focus different sets of consumers or individuals. Targeted lists are generated that includes past customers who present higher chances to respond to specific messages. Hence, email marketing gives high ROI to the campaign managers.

Variety of tools, techniques, and efforts are required to enhance the pace of email marketing further.

Different kinds of email marketing

- **Direct emails**

As the name suggests, these types of emails provide direct information regarding a product, promotional offers or sales. Customers receive these messages along with a call-to-action and links. An interested customer then uses the connection to obtain products, coupons or other services.

- **Newsletters**

Many customers subscribe to news, information and other messages. Companies send regular emails to such customers in the form of newsletters. Email newsletters enhance relationship building process to improve brand loyalty and customer retention. Marketers create messages in a conversational format to engage audiences towards the provided news.

- **Transactional Emails**

These messages indulge customers in some kind of action. A customer can provide purchase confirmation or trigger some other transactional action. Marketers generally blend a sales pitch with transactional emails to engage customers in other products or services.

The Benefits of Email Marketing

A correct approach to email marketing can help you reach thousands of consumers on a weekly basis. It is all about creating a system in the beginning. Eventually, you become capable of sending thousands of messages consistently. Thus, it provides with multiple advantages such as:

A targeted list of customers

If you follow a systematic approach, email marketing always provides you relevant traffic. Relevant traffic means that you send messages to customers who are most likely to buy a product or services you are promoting.

The process of email marketing starts with list creation. You generate multiple lists based on demographics that belong to product niche you have in mind. Your content, sales pitches, product promotions and other messages attain a directed pathway.

Hence, it would not be wrong to say that you never waste your time when it comes to email marketing. Relevancy lies at the core of this marketing method.

Automated marketing process

With advanced tools present, different steps of email marketing have become entirely automated. Email creation, bulk delivery, and follow-up, everything is possible with a few clicks these days. Hence, you don't have to spend too much time managing everything.

Many tools help you create customized messages. You can update message creation processes and design a sequence around your email marketing strategy. For instance, you can define daily, weekly or monthly emails for the list you create.

Different automation processes are leveraged to create brand awareness. Then, you can move forward and design engaging messages for more interested customers.

Cost-effective

The return on investment is so high that email marketing attracts all kinds of businesses and individual marketers. You already know how automation empowers your ability to send emails without wasting any time. The same system also allows you to save a lot of money in the process.

You don't have to attain a marketing certification to start earning through email marketing. It eventually concludes with a few simple steps and the organized approach you generate. Then, you can receive a great deal of money annually.

It generally requires less than $10 per month to start your marketing process. You can easily register with email sites and use collected email IDs to send messages to consumers.

If you think it through, less than 10 dollars will allow you to create an automated money earning email marketing process. And that cycle will exist forever and let you gain hundreds of dollars in return. There is no better method of creating this passive source of money, with such a small investment.

Personalized engagement

In any form of marketing, personalized messages are considered most effective. However, marketers usually worry about the hard work required sending personalized messages to thousands of consumers.

Personalized engagement is not difficult nowadays. Most servers provide a customizable system that allows you to include receiver's name automatically.

For example, your message can start with:

Hey, (receiver's name), hope you're doing well.

Adding the name creates a personalized engagement, which enhances the chances of a response. The receiver feels connected to the message and opens an email to go through the content written inside.

With the right tools, you have to spend a few seconds to design a personalized email sequence and send messages to your consumers.

Simple to create

Message creation is effortless in email marketing. You don't need any technical knowledge or a big team by your side. Appealing templates, images, videos and other media resources are readily available in the market. You can collect these media on your own to create beautiful content. However, many marketers use text emails successfully with minimal media to engage people.

In emails, content is the most crucial part. All your receivers desire information that is valuable and useful. Hence, you can utilize everyday text to generate valuable messages. You don't even have to be technically savvy. There are tools in the market that allow you to drag and drop to design your emails in simple steps.

A few minutes in front of your laptop is enough to generate a sales funnel that works on its own. In return, you create a source of income for your whole life.

Easily measurable

For a marketer, it is a necessity to stay on the right track. Your whole business cycle depends on how directed your marketing strategy is. Email marketing processes are highly beneficial for tracking. Advanced software present valuable data associated with a campaign's success rate.

You can easily find out about impressions, CTR or click through rate and conversions as well. These are the most important data structures that help you monitor your email marketing approach. Changes in data allow you to take immediate actions and improve your strategies for better results.

The definition and workability of analytics is fathomable even for a person who is not a data scientist. These are simple numbers that come with a well-defined format. Most tools give a visualized form of data, so that, anyone can understand it.

Eventually, you become advanced and start reading analytics faster. You can align your subscribers' growth with click-through rates obtained daily.

Similarly, you can choose tools that help you compare conversions based on location, time, day and other factors. Having all data visible at one place saves hours of effort.

Highly scalable

No other marketing method connects you to a global market. Sending emails to thousands of audiences all over the world. Social media is also there with similar reach. However, there is no guarantee whether your audience is reading your message or not.

Email Marketing offers the much-needed scalability to enhance your earnings with time. There is entirely no limit to how much you can grow with email marketing. This massive scalability attracts people from all backgrounds.

Your list keeps on growing, and that also improves your ability to make money on a grand scale. It is true that you need show patience and determination to attain more than 1 million subscribers. But that is not too difficult if you design a proven approach and stay consistent. A single individual can generate a massive list of emails within 12 months.

After generating your list, you have to sit back and let the system work for you. One day of work completes your task and creates a cycle of immediate income and high returns.

There are several tricks to keep improving the scalability of your email marketing:

- Regularly verify email addresses available on your list. Remove and replace the ones that are not working. Accuracy is the key to consistent scalability.

- Separate your messages from other promotional emails. Design a personalized message for every recipient. It will increase CTR and also positively impact conversion rates.

- Emojis are highly popular these days. You can incorporate them to make your message look more fun and entertaining. The subject line is a great spot to place emojis to stand out quickly.

- Experiment with different delivery times and compare impressions and conversions. This way, you can modify your system for maximum open rates.

- You can even try creating a separate list of people who are more responsive to your messages. It is a great approach to design your marketing messages around those confirm sales.
- Customize templated for a unique appearance. Make sure that those templates are responsive, as most people open their emails on their smartphones.
- Your email needs to have simple-to-understand CTA. Your subscribers shouldn't feel confused about what you desire from them. Short emails followed by a clear CTA are perfect to get more responses.
- Give yourself time and let your list grow. Initial stages ask for a little bit of work. But then, you can see regular growth in responses from recipients.

Huge profits

An email marketer receives vast earnings with minimal efforts. If you carefully look at all the benefits as mentioned above, they all point towards passive income. Email marketing enables you to create lucrative income on a regular basis.

Think about the investments required for implementation. As mentioned earlier, you have to spend a minimum amount every month. The implementation process is extraordinarily cheap and takes no effort. So, you feel safe regarding investment in this passive income business.

Now, the gradual growth of your email list keeps offering a steady rise in your income. Eventually, you reach a large size of the list that takes your income to an unimaginable scale. You can expect profits up to $1000 every month if new subscribers keep on coming.

With an automated process, it is possible to add more than 500 subscribers every week. These subscribers increase the number of sales you make. As a result, your overall monthly income improves. It is all happens with $10 to $50 investment every month.

There are a variety of ways to leverage your subscribers for profits:

- Sell your product or service such as training guides, eBooks, online courses, virtual coaching or other services.
- Generate an affiliate connection to sell products or services on behalf of other businesses.
- Provide exclusive products to people who are interested in that product category.
- Modify your offers based on the interests of customers you have on your email list.
- Monetize streamlined messages to get multiple sales from customers. Seasonal and perishable products increase your chances of making repeated sales.
- Send follow-up messages requesting word-of-mouth from your subscribers.

If you can create an appealing email copy, all these methods are excellent to achieve more profits through email marketing. Most importantly, study your subscribers and their buying behavior. You can use these insights to write more effective email copies and generate more profits.

How to build an Email List?

Till now, it has become visible that the impact of your email marketing system depends on the email list. A great list of email IDs includes correct information regarding target market.

Five steps of creating an email list

Here are five most valuable steps for building an email list for passive income.

Step-1 Finding your market

To make money, you have to find a market first. Most people create their business through blogging platforms. The niche of your blog can give you a direction and help you generate your business prospect where you lead as an expert.

Just like any other market, an online market also demands specific traits from a marketer. It is your responsibility to understand those needs. You can't just create a list of random people with their email IDs. It is necessary that they are a consumer of your products. Hence, you should narrow down your market based on your expertise, your products, and services.

Step-2 Knowing your market

You need an in-depth knowledge of your market to reach out to them in an effective manner genuinely. Your list of subscribers continuously grows when you understand and act according to what your market wants.

When understanding your market, focus majorly on their buying behavior. For instance, some niches attract more customers with low-priced products. However, in other niches, you can easily convince people to pay high prices for a product. It all comes down to how that market works and the purpose of subscribers when they opt for your list.

Step-3 Generating your automated response sequence

The conclusions attained through market analysis will help you define your product. This product can be an incentive, discount, content or any other item to attract people. Now, the next thing you need to worry about is creating an automated sequence for the response.

An automated response sequence includes a series of emails that you design for subscribers. These emails should contain valuable messages and free information followed by promotions.

Email list building is not just about gaining email IDs. You have to present your emails as a valuable source of services in the initial stages. Then, you can pitch promotions and get conversions.

Remember, that you don't want free service seekers in your list. Hence, blend your free content or services with promotions. This will prepare your subscribers to become a customer.

Step-4 Placing an email signup landing page

You need a robust page that links your market to your email facility. This page is called a landing page that you place on your blog or site. A landing page provides a bright idea of what you need from your visitors. However, a successful implementation requires following qualities:

- Create a headline that appeals to the visitor.
- Choose benefit-oriented adjectives in your headline.

- Describe subscription benefits in detail via short bullet points.
- Crisply present your product with a short description.
- End with a simple call-to-action.

Step-5 Promote your email signup page

You have online as well as offline methods of promoting your email signup page. You already understand your target market. So, use that knowledge to market your email signup page through social media, blogging, forum posting, direct mailing, community events and other methods.

Tips when generating your email list

Here are all the right ways to build your email list:

- **Connect through your blog**

A blog is a great choice to create a huge community online. In fact, you can directly connect with potential customers and audiences on a regular basis. Plus, it offers you an opportunity to build a substantial email list for your target market.

Consistent blogging with a clear call-to-action will make your readers more interested in signing up for email messages. An active blog can help you create a massive list of emails in a short period.

- **Gather from social media**

Social media platforms provide a huge market where you can connect and build relationships. These interactions allow you to find new audiences who might be interested in your product or service.

You can offer solutions to their problems and encourage them to sign up for your answers. These emails will be a great addition to your targeted list.

- **Leverage telemarketing**

If your business or job allows you to interact with many people, it is an excellent source of emails for you. Whenever you call a customer, ask if he or she would prefer getting personalized messages via email. You can provide information regarding your goals with the email list. However, focus on customer-oriented goals during this discussion. Tell them, how joining your email list would help them on a regular basis. This will enhance the chances of getting more and more subscribers.

- **Ask for business cards**

You regularly meet professionals who seem like they would prefer your services. Make sure that you get their business cards and exchange it with yours. These cards will help you create a list of emails that belong to highly targeted prospects.

If you have an office, ask your visitors to leave their business cards. You can give some incentive to boost card collections.

- **Get a software**

Many email list building tools are available out there. Good software allows you to find email data targeted to your service or product directly. You can start by researching multiple tools and choose one that matches your needs.

- **Offer coupons or discounts**

People are more likely to join your email list if they get valuable offers in return. And nothing seems more valuable than a discount coupon. Almost every online business model leverages this technique. You can provide discounts in return for joining your email list. This will add a considerable number of people on your list.

- **Promise to deliver free content**

Free content is another reason why many people like to join. This is an affordable alternative to discount coupons. However, it is essential that you provide a small piece of content only. But this content needs to be valuable and directly associated with your products and customers. You can create articles, ebooks, videos and other types of content pieces for your email list.

- **Create an engaging form on your site**

A site or blog can have a separate page or popup page including sign up forms. You can place this form as a link right in the navigation bar on your site. Or, add this to your site menu.

- **Give email specials to your online community**

With blogs and sites, you can offer content that engages a vast population. But that keeps your community limited to your platform.

And as you start getting people to sign up for your email list, you start offering them email specials. These specials are services, content, products or discounts that will be available to your email subscribers only. With this, you can encourage your subscribers to share your services to others as well.

- **Reply to blog or site comments**

Two-way communication through your site or blog is way more useful for gaining email subscribers. You should answer questions on blogs, forums, and websites. Do it on your blog or site along with other sites that allow commenting. Q&A forums are highly active for interactions. You provide a solution to a problem and get a chance to pitch your email service to prospects.

- **Let word-of-mouth work for you**

A continuous cycle of email subscribers becomes possible when you leverage your current subscribers. Every time you send an email, include a request for sharing services. Your subscribers will share your services with their family and friends. Hence, you will get more emails on your list.

Remember, that your mail should contain a sign-up form. This way, every recipient of that mail gets a chance to opt into your emails.

- **Charm with multiple subscription choices**

More choices make people more interested. And you can use this to your email list advantage. You can create a variety of content associated with your product or service. Then, design multiple subscription options for all different content categories. This way, you can charm site visitors by offering them multiple choices. They can choose to subscribe for one of your content categories and gain regular emails that are relevant to their needs.

Along with the choices, you can also give freedom to your subscribers concerning email frequency. Ask them whether they like to have emails on a weekly basis, monthly basis or every time you have something new to offer. It is a great approach to give a sense of control to your audiences.

- **Give away exclusive courses on your expertise**

Fitness trainers, teachers, financial advisors, all experts have a great deal of knowledge gained through their experience. If you have such expertise, leverage it to enhance your email list.

Create a free course around your expertise. You can offer a weight loss course, cooking class, or any other. Then, start promoting this course through your blog, site, social media, forums and other online platforms. Make sure that you provide this free course only via email subscription. Hence, a person will get your particular course only when he or she subscribes to your list.

Segmenting Email Lists

Segmentation of email lists is a well-known technique used by email marketers to define highly target groups. The process includes dividing an email list into multiple segments. As an email marketer, you should find out specific needs of your subscribers. This understanding helps to choose right groups for right email campaigns. Hence, you get more conversions through your email distributions.

Best practices for segmenting email lists

Every email marketer sells a unique product, service or idea which is why everyone needs a particular segment of the market to cater. However, the pathway to segmentation goes through similar points. Here are major steps that you should incorporate to segment your email lists successfully:

Email list segmentation tool

Thousands of emails on your list can seem difficult to segment manually. So, finding an email list segmentation tool seems a logical move. You can choose among various email service providers by looking at the segmentation service they offer.

New email subscribers

Your new subscribers don't know your services. Hence, you can't approach them with the same system you follow for the old ones. It is important to offer them valuable messages to prepare as a customer. You should segment new subscribers and generate a separate email approach to convert them into customers. You can start by sending a welcome message along with a short plan of what you have for the subscribers in coming weeks. Then, keep sending valuable emails with resources that match your subscribers' need.

Interest segmentation

With a reliable segmentation tool, you can have a huge dataset associated with subscribers' interests. These interests prove a great segmentation approach to enhance conversions. For instance, you can offer weight loss products to subscribers who have a great interest in the fitness-related content. Or, provide email fitness courses to the segmented group.

Service preference segmentation

Your email list includes your customers. And they need to feel happy with your email services. Understanding customer preference is highly important to keep them happy. In email lists, customer preference includes types of content, the frequency of emails, discounts, and other updates. You should describe different segments of subscribers according to their preferences. This will save you from losing subscribers.

Rate of engagement

In your huge email list, you will experience that there are people who consistently engage with your messages. However, you will also find a group that rarely opens. Your software will tell you about the open rate. You need to segment subscribers into three categories.

- High open rate
- Medium open rate
- Low open rate

This segmentation allows you to know which users can be pushed towards high rate. You can present rewards for medium open rate subscribers to gain more engagement from them.

Similarly, you can generate an email plan for low open rate subscribers too. These are inactive ones, so you need to remind them about your services. For instance, if you offer exclusive courses, send reminders to people who are not following weekly course resources.

Lead segmentation

When signing up, every subscriber opts for different lead magnets. These lead magnets are the type of services or products they desire. Lead segmentation will simplify the process of messaging and promotion. You can create a group of subscribers who desire PDFs, webinars, downloadable articles or other types of content. Thanks to automated tools, this process won't take too long.

Opened, but not converted

You will always find a difference between open rate and conversion rate. Many people open emails, go through the message but don't follow the conversion process. You can create a segment of such subscribers and let them know about the missed-out opportunity.

You can start by sending a detailed message with valuable information related to your product. It is possible that those people weren't satisfied with the previously delivered information. So, you can introduce your product in a detailed manner to get more conversions.

Purchase history

This segmentation is highly essential for marketers who promote multiple products via emails. Some people buy same products again and again, while others choose different products every time you promote. Knowing purchase history will help you reduce your efforts and improve results. You will know which list member would be ready to buy your product. This will save from making unsuccessful efforts.

Buyer's cycle and frequency

People who regularly buy from you should get rewarded. This brings loyalty to your email marketing. You can send an invitation to your frequent buyers to join new campaigns and get benefited.

Similarly, you will find higher conversions in different time intervals. This is called buyer's cycle, which depends on your product and buyer's need.

When marketing through emails, you can design a message delivery system that matches your buyers' cycle. This increases the chances of conversions, as you present your offer right when your subscriber needs them.

Desktop and mobile customers

Most email marketers don't realize, but this market segmentation is extremely important. An email looks completely different on mobile due to smaller screen size. It is prudent to fathom the number of subscribers, who read your emails through smartphones. Then, you can design email copies for high readability and target accordingly.

Try to find responsive email templates that look appealing on a desktop as well as mobile.

Subscribers who help in promotion

As you grow your email list, your subscribers start referring your email services to others. You should take special care of these subscribers. You can send a "thank you" message along with rewards regarding free content. This will encourage them to keep on sharing your content with others and grow your email list.

The profession of your subscribers

His or her career majorly defines a person's interests. A content writer looks for content resources. Similarly, a physical trainer looks out for offers on fitness products and equipment. Understanding and segmenting your subscribers concerning their profession is a great move. You can find different selling points and address problems with your products and services as a solution.

Gender segmentation

If your marketing strategy is directed separately towards men and women, you need this segmentation. Define gender groups in your email lists and design focused messages for both men and women.

Segmenting for education level

For some email marketers, education level of a subscriber matters. For example, if you desire to sell multiple courses on marketing, you should know the education level to send right emails. You can't expect a beginner marketer to respond to an advanced level of course. Hence, it is essential to divide your email lists into different education level categories and send course-related emails that are most suitable for a group of subscribers.

Segmenting according to income

Your subscribers choose products and services according to their buying capacity. You can understand that buying capacity by knowing their average income. Define different ranges of income category and put email subscribers in those categories. You can even go up to 5 figures or 6 figures incomes, depending on what kind of product you want to market.

Every member of your email list behaves differently. It is your job to understand those behaviors and modify your marketing approach time to time. Segmentation through automated and manual processes can unlock the maximum potential of passive income through email marketing.

Qualifying your Subscribers

Before you begin your message distribution, make sure that your subscribers want your services. Qualifying emails means that you inform your contacts about your message distribution plan. This way, you don't send emails against any subscriber's will and don't seem like spam.

There are many different ways to qualify your email list and ensure that they are your opt-in subscribers.

How to qualify your subscribers?

Is your email list legitimate and qualified for marketing?! That depends on the methods you leverage to obtain your list.

Site visitors who permitted you to send emails

These are your opt-in email list members. They know about email services at the time of accepting your services. They sign up for your emails by giving their permission on your site. However, it is also vital that you clearly state what you want to do with their emails; if you wish to send newsletters or desire to include them in marketing campaigns.

It is advisable to confirm such subscribers by sending a confirmation message. This will ensure that they qualify as your opt-in subscribers.

Subscribers who filled a form on your landing page

If a subscriber fills all the information asked on a landing page, he or she qualify as a subscriber. But the form should explicitly say that you are asking for emails for campaigns. There are automated tools you can use to design your landing page and get subscribers.

Such subscribers usually know and understand the services. Hence, you can include them in your opt-in collection. Some marketers like to get a second confirmation by sending a clear email including the details of campaigns you plan to run.

Business cards you obtain from office visitors and trade events

Generally, these people don't know if they are on your email list. So, you can't count them in your opt-in collection. The right way to incorporate these contacts is to send a series of emails in a non-pushy way. Describe what you are trying to accomplish and benefits that you can offer. Then, it is up to those people whether they accept or not. Keep adding people who confirm the acceptance of emails.

Members of your social media community or forum

Sure, these individuals are more likely to accept your emails. But you can't include them in qualifying list without getting their permission. A subscriber should know and have the necessary understanding of your campaigns. You need to contact these members and ask whether they would like to become a subscriber or not. Try posting a link to your landing page and direct these people to the subscription form. This will make them a legitimate contact.

Customers who have bought a product or a service from you

Depends on how they bought your product or services. If the medium was email, then, you can count them in your opt-in collection. However, products sales via social media, site, or any other method requires follow-up. You can suggest customers that they can get more services or products time to time via your email services. Get their permission to use their email IDs. Then, you can keep sending them new pitches.

Emails available in your address book

During list building, you can include all contacts available in your address book. However, they don't necessarily qualify as your subscribers. Send your marketing proposal in an email and ask for confirmation. If they accept, then, you can include them in your opt-in collection.

The list you have rented or purchased

If you have rented a list from another email marketer, they don't qualify as your subscribers. You can use them to increase your email list. But it is critically important that you treat them as potential subscribers. Send valuable messages to engage them. Then, offer content useful to pitch your products finally.

Clean your email list

After qualifying your subscribers, you can follow a few necessary steps to clean your list. Cleaning your email list includes different methods to enhance the quality of emails.

- **Get rid of misspellings**

First of all, you need to resolve misspelling and typos. Collect email IDs that bounce back whenever you send messages. Review these IDs to find typos and resolve them.

Also, check which emails don exist any longer. These IDs have nothing to offer, so it is wise to remove and try to look for new emails from the same subscribers or others.

- **Find inactive emails**

Sometimes, correct email IDs stay inactive too. You can find these emails when checking data associated with an open rate on a monthly basis. This data will allow you to gain insights about inactive subscribers. People who don't even open your email for months require special attention. You can create a special campaign for such subscribers and ask for their needs and interests.

For inactive emails, you need to create an appealing subject title to grab their attention. An offer or a discount can help in this process too. It will turn many inactive emails into active ones again.

Building a Relationship with Your List

An essential aspect of email advertising is building a healthy relationship. Sure, you get a better response via emails, but that relies on how you connect with your subscribers. This is a fast-paced marketing world. Things move fast, and people don't have too much time to offer to any random marketer.

As an email marketer, it is wise to use high-quality technologies. However, you should keep the basics of marketing in mind. Finding a potential customer is one thing but keeping that customer on a regular basis requires consistent efforts.

Whenever thinking about relationship building, you should concentrate on three major factors- **awareness, likability, and trust.** Achieve these factors, and you become a successful email marketer.

How can you achieve all three factors?!

Start with message personalization. A personalized email connects you directly with your email list. Tapping into email marketing is a success only when you create strong relationships with your actions. Hence, utilize all segmented data associated with education, location, gender, and others to make your messages more personalized. Keep an eye on consistent and rare open rates. Define subscriber-oriented message strategy. That is how you win every subscriber and develop a strong, long-term relationship.

Become a consistent and disciplined email marketer

Your relationship with email subscribers is similar to your general social relations. You regular spend time and interact with your friends and family via calls, online and offline communication. The same approach helps in email marketing as well.

Your email list will grow concerning loyalty if you stay connected. Your emails should reach your list consistently. You can define a disciplined system on a weekly and monthly basis.

However, the key is to stay stick to that system. Only then, your messages will receive a high level of attention.

Share a variety of content that matches subscribers' interests

Your subscribers should trust you as an expert resource. So, you need to become a one-stop solution for relevant content. Combining text messages with appealing media, videos, audio and other sorts of content would be a great marketing move. However, they all work when you only keep the interest of your email list in mind. Choose different ways to present relevant information. Eventually, your email subscribers will start looking at you as a reliable solution. Then, you can think about product promotions and selling items.

Make them click every email with a dazzling headline

You can connect to your email list only when they open your emails. And that majorly depends on how you write your message headlines. A dazzling subject looks different than other messages in an inbox. It attracts people with a clear, crisp and beneficial message.

Here are a few examples to create click-worthy subject lines:

- Use brackets to highlight special features of your emails. You can write [FREE EBOOK] or [VIDEO CONTENT] to grab the attention of your subscribers.
- Use emojis and numbers to write more attention-grabbing subject line.
- Present your subject as a question to generate curiosity.

You should consistently experiment with new types of subject lines. This will evolve your techniques to gain higher open rates.

Optimize emails for user convenience

Email users differ regarding priorities, platforms, and devices. Your job is to understand those differences and optimize your messages accordingly. As mentioned earlier, optimization for mobile devices is the most important part. Maximum conversions come through mobile users. When you think about your users, they start responding more. Make emails convenient to read on mobile, and you will see better loyalty rate from your email list.

Many users don't allow images on their emails. You can include ALT text in your images for such users. So, even without the images, your user will see the name of that image along with a clickable link. Similarly, you should define goals of your subscribers and incorporate them in your email copy. This will reduce confusion at the time of message production. Plus, you will deliver exactly what your users want to read.

Don't just try to sell

If you try to sell things, soon your subscribers will stop opening your emails. The whole idea of finding and segmenting email lists is to provide value to your target market. You should always divide your email campaign into two parts. One part should cover content and resources to help audiences. And the rest part should focus on selling products or services. Smart marketers follow 80 and 20 percent parameter when dividing marketing approach. Larger part goes to providing valuable resources.

With this approach, you connect with your customers on a personal level. A relationship generates concerning mutual interests. And that is how you sell more in future. Regular valuable resources make your promotional pitches more effective.

Subscribers pay attention to your emails and respond by buying those products.

Combine multiple affiliate deals in emails

If segmented smartly, you can promote numerous affiliate deals in one email. This improves the chances of boosting your profitability. But it also presents you as a reliable source in front of your email list.

Choosing the right affiliate deals is the key here. You should combine non-competitive products, so that, they don't cancel each other's solutions. Focus on your content first. Point out different problems through your email content. Then, present complementing affiliate offers to resolve those problems. Just remember that those deals are desirable and match your target market's need.

Write interactive call-to-actions

The most significant mistake that marketers make is following a formal call-to-action format. Multiple lines, in the end, asking for different actions make you look too profit-oriented. The end of your email should have an interactive touch. The call-to-action should include everything you need from your subscribers. However, the writing format needs to have a light communicative model.

For example, you can write, "A great course is waiting for you on the other side, click now to start growing."

This line includes all standard features of a call-to-action:

- A link to click to
- A crisp sentence
- A pathway towards making that purchase

But that is not all! You can see how light and interactive this call-to-action is. And this property further enhances the chances of getting conversions on a regular basis.

Use impressive rewards

Almost every email marketer tries to impress people with discounts and rewards. Subscribers receive hundreds of emails with such offers. But you can create loyalty in your email list with unique and relevant rewards.

Always think about your segments' interest before deciding rewards. For instance, long-term subscribers would love to get rewarded for their loyalty. Similarly, you can present special rewards for word-of-mouth for those who promote your campaigns.

Always publish on-time

Email marketing allows you to define your approach to message delivery. However, that doesn't mean you should follow a random publishing system. Your subscribers should be able to put their faith in you. There are some marketers who lose their email list due to not providing resources on time.

You can design your email distribution frequency according to segments. But then, make sure that those messages reach subscribers in the same manner. Your audience will be aware of the fact that 'when to expect your emails,' which will positively enhance your open rates and conversions.

Thanks to email distribution tools, you should feel any difficulty in following a disciplined format of publishing.

Avoid sending too many emails every day

If you want them to respect your marketing approach, you have to respect their time as well. Ultimately, your email list is a list of people who have their own lives and daily responsibilities.

It is critically important for you to recognize that and respect that. Don't send too many messages on a daily or weekly basis. A constant reminder of the same thing can cost you many subscribers.

Design a balanced system that fulfills your marketing plan and keeps your subscribers happy too. A maximum of 2 times a week is a good system to follow; and increasing your frequency for those who engage in your emails.

Always send easily consumable information-packed emails

This is another aspect of not wasting your subscribers' time. You need them to take your messages seriously. And that won't happen if you keep on sending unnecessary messages with too long bulky content.

Information is important. However, a smart presentation of that information is much more valuable for email marketers. A person gives a few seconds to an email, so you need to point out very effective components to make an email work. Write an email copy that enables a quick scan through naked eyes.

Short sentences, bullet points, crisp concepts, these are the elements you need in your format. It gives a whole lot of information in a very short time period. Hence, your subscribers feel more comfortable opening and reading your emails whenever you send.

Choose your autoresponders wisely

An autoresponder saves from spending all your time replying to new subscribers. It immediately sends an automated message to new people who get added to your list. Make sure that you choose a reliable and swift autoresponder. Also, ensure that your automated responses are warm and engaging. This will create a strong first impression and set a foundation for future communication.

Choose a casual tone of conversation when designing such emails. The language should be welcoming and appropriate as well. The idea presents you as a friendly and credible source. Then, you can include this email in your response system.

Monitor results of your efforts and modify

A strong relationship is not a one-day job. You have to regularly improve your abilities and interact with subscribers to gain their trust. Consistent measurements can help you do that. Keep an eye on multiple email metrics to understand how engaging your efforts are. Give additional focus to open rate, CTR, and conversions. This will offer insights associated with trending interests, most successful segments, and high-rewarding campaigns.

Evaluating data regularly will set a foundation for relationship management. Then, you can use these insights to generate different scenarios. Try a different approach to see how your email list responds and improve accordingly. Make sure you include sales and leads when estimating your overall ROI.

Focusing on every step mentioned above, you can chase your profit goals without losing the trust of your email subscribers. Your emails will become appealing for all kinds of segments no matter which device they use. Consistency and balanced frequency will obtain you an undying curiosity. Hence, you will develop a regular communication channel with your messages.

Make Money with Your List

Here are some unique ways to make money with your well-segmented email list.

Indulge in affiliate marketing

There are many brands and businesses out there offering email marketing affiliate commissions. Such affiliate programs allow you to market products and gain commissions in return. This approach is highly popular for passive income generation. Any person can create his or her email list and incorporate affiliate programs to make money. You can easily earn up to 25 percent of overall product price commission. A simple inclusion of affiliate link in your emails is enough to let the magic happen.

Marketing a product is easy. The real critical part is selecting products for affiliate marketing. You need to find products that are relevant to the interest of your subscribers. Plus, those products should have affiliate programs too. Brand power also matters if you come across two options in the same product category.

With in-depth analysis, you can choose a perfect product for affiliate marketing.

Offer advertising space in your emails

When you successfully build a large email list, it is possible to run ads on your newsletters. You need to have more than 2500 subscribers to make it happen. Email marketers charge great amounts to provide a small space for ads.

An email with advertisement includes several lines associated with that product along with a link.

Sell your products

You can sell your products such as courses, books or others, depending on your expertise. The best approach to selling would be presenting your product to your subscribers. Promoting your product is not about commissions. You need to try and create a continuous cycle of sales, which is a little harder in this case.

First, start by clearly defining what your users want. List segmentation will help you do that. Plus, you can directly ask your users about what sort of products they would like to have.

Direct their attention towards a problem and present suitable solutions. This will allow you to understand which solution is most trending among your users.

Use answers to design your products in a mold that impresses the most. It will surely take plenty of time, and in some cases, money as well. You can take over the whole product generation process to save from high expense.

When you are designing your product, remember that quality is the most important factor. It doesn't matter if it takes longer, you should never compromise on quality. Pay attention to details and regularly match product's growth with market needs.

Finally, you can start using relationship building techniques to promote your product.

Rent email lists to others

Not very common, but many email marketers use this technique to earn. You can get quick money by renting your email lists to other marketers and businesses. However, it all comes down to your preferences. Most marketers spend months in generating and optimizing their lists. So, you may desire to keep that list for your marketing goals only. But if you do choose to rent, make sure that you rent to a legitimate and credible source.

Follow a proven promotion path

As you keep trying multiple ways to lock-in customers, some campaigns will thrive, while others won't. A campaign that has given you positive results in the past is worthy of a replay. You can gain the same excitement and sales with repeated email marketing campaigns.

The process is simple. You need to keep an eye on campaign performances. Notice sales and ROI of all your campaigns and align them in descending order of success. This list will help you choose which campaigns are perfect to try again. A product with discounts or special promotions can make people pay instantly. So, you shouldn't stop yourself from trying proven methods again and again.

However, make sure you don't make your campaigns look too repetitive. Balance in such a manner that your subscribers look at your campaigns as a fresh deal. And that will bring you positive results every time you try an old campaign.

Match sales approach with list segmentation

You already know how campaign segmentation can skyrocket your marketing ability. But it depends on how well you choose products for every segment. Segments associated with open rates, demographics, past purchases, product interests and others are important here. You can use these segments to define pitch force and type of products. Email subscribers will see products that they look for. Hence, the conversion rate will increase for sure.

Use an up-selling approach

Up-selling is when you top one product with other at the time of selling. An email subscriber accepts a product, and then, you offer another complementary product to enhance the price. It is essential for the second product to work as an upgrade for the first one you have sold.

Think about how a restaurant service asks you for beverages when you order a burger. Up-selling is the same. In your case, second products can be upgrades or even additional services. The product can change with industries, but they all come under the concept of upselling.

Use down-selling approach

Sometimes customers accept then back out without purchasing the product. This is common, as people feel uncertain about their investment in products. Your job is to make that deal more valuable for them to get them to buy that product.

If possible, try to lower the price, so that, they can buy. This process is called down-selling. Make sure you make that reduced price look like a one-time offer. This will create a sense of excitement and increase the chances of conversion.

Re-target "opened, but not converted" segment

Most of your marketing efforts can face low conversion rates due to this problem. However, you are not the only one. Every online business has to fight this problem. You can copy their approach in your email marketing.

People who open your emails, but don't buy have some issues with your product. Your job is to resolve those issues by offering a series of valuable emails. This is called re-targeting. You provide detailed product information, present discount opportunities and remind them about the benefits of the products.

Re-targeting can increase your conversion rate up to 50%. Campaign's CTR will increase, and you will see a boost in sales as well.

Sell premium content through a paid subscription

Do you have expert knowledge in your field?! Then, your email list can become your market. You can develop a vast collection of premium content and sell them through paid subscription.

To attract subscribers, you need to provide them free content. Offer valuable content pieces and include a call-to-action informing them about the paid subscription. You can provide one-time content, monthly or even annual membership to your subscribers. Dividing into different formats will allow most of your subscribers to choose paid content, according to their preference.

Even with a small payment, you create a consistent cycle of revenue. Your content is there forever, so all you need is new subscribers to increase your sales.

Start one-on-one consulting or coaching

Email marketing is a great way to build a strong consumer base. Your consistent help and suggestions offer benefits to subscribers. As a result, customers start relying on your content. Then, you can pitch your coaching or consulting services. These services usually include having one-on-one interaction with subscribers. That is why you get paid way more than content selling.

Offering to coach via social media or any other platform doesn't give you that much exposure. However, emails create a sense of personalization and security. Hence, people are more likely to connect to get your help on a regular basis. You can divide your sessions according to your free time and make money.

It is necessary to remember the importance of quality here. Your suggestions need to be effective and practical. A customer should be able to use your coaching to generate desired results. Only then, you can expect continued growth in your market base.

Choose your coaching content very carefully. You should have experience and effective knowledge in that field.

Rules that will help you make money consistently through your email list

Till now, you have received multiple methods of making money with email marketing. But the success of those methods is possible when you follow certain rules:

Rule#1 Pick your niche and stick to it

No business can achieve success without picking a niche. Your customers should know what you are selling. Hence, you need a niche. Trying to pull in customers in different categories can limit your ability to grow. It usually confuses your subscribers, and you don't get to define your marketing approach.

There is nothing wrong with selling multiple products. However, all those products should belong to the same line. For example, you can sell cooking recipes, cooking appliances, and cooking coaching classes altogether. The idea is to target a certain market and segment according to their needs.

Selecting a niche makes things easier for you and your subscribers as well. People feel clear when they decide to sign up as a subscriber. Otherwise, your services seem all over the place, making people think twice before opting for products or services.

It is critical that you choose your niche according to your email list. Hence, the process of niche selection should start way before list building and segmentation. This way, you become familiar with what your market needs. Then, align those needs with your expertise and apply in list building.

Rule#2 Make your pitch valuable

You have read the term "valuable pitch" for many times in this book. But what does that mean?! A valuable pitch matches the needs of subscribers in a unique manner. Marketers who make money don't write their sales pitch focused towards product features.

To write a valuable pitch, you need to understand the **difference between product features and product benefits.** A potential customer responds to product benefits, rather than product features.

You can't focus too much on your product. Of course, your purpose should be to sell your product. But that is possible when you present that product as a solution. People think about their problems whenever they see a product. If they don't, conversion doesn't happen. Your job is to remind about a problem and present solution through your pitch. That is how you can make your pitch valuable.

Rule#3 Define a path towards sales

A content piece should not just be a content piece. Every email, message, and resource you provide to your subscribers should align in one path. And that path should lead to sales.

With blogs, articles and other content, you can make your services valuable. So valuable that your subscribers start waiting for your emails. Then, you start pitching and upgrading your emails from resources to sales pitch. Inviting subscribers to buy a paid product, then, works effectively. Hence, you can sell your eBooks, courses, items and other products.

All in all, every content piece should have a purpose. So, it is not just about enhancing the value. It is about enhancing the value in a certain direction. That is how you make money through email lists.

Many marketers become pushy and send spam emails. This happens when a marketer doesn't define a clear email sequence. Overdoing pushes customers away. You have to stay away to increase email sales. Treat your subscribers as your friends and provide with genuine offers via your email sequence. Present your offers in multiple manners without overdoing it.

How to convert every subscriber into a customer?

You complete 50% of the job by finding the right emails to add to your list. However, the rest of the job relies on your ability to turn those subscribers into customers. You already know various ways to make money via email marketing. But it is also necessary to ensure that every subscriber becomes a customer. This approach will lead you to maximize revenue.

Here is what you should do:

1. A well-defined sequence

A successful email sequence includes following steps:

- An email that congratulates subscribers for the subscription
- An email that tells stories of happy subscribers.
- An email that offers products along with links and CTA.
- An email that presents deadlines to purchase products for maximum benefit.
- An email that reminds final moments to grab an amazing deal.

You can incorporate an autoresponder to create such a system. There are many paid tools available that you can find and leverage.

2. Make buying easiest through your service

There are all kinds of products available on the internet. So, the question arises- why a person should buy products from you?!

You can present speed of purchase as your answer. Of course, the quality of products or a service matters. But it is necessary that your subscribers obtain a smooth buying experience. Hence, you need to be precise when creating your sales emails. Also, optimize those emails for all kinds of devices.

Try to include as few redirections as possible and provide a common payment model. This will help customers to buy from you on a regular basis.

Conclusion

As an email marketer, your email list becomes the best asset you can have for passive income. It is your list that helps in building relationships, finding regular customers and making money on a regular basis.

No matter how many social media platforms are launched, emails are going to be valuable forever. However, just building a list of emails can't provide value. You need to find the right types of subscribers to make marketing profitable. So, your major goal is to remove irrelevance from your list. Then, the rest of your subscribers will become your long-term customers.

Keep on reminding why you are building your email list. You ultimately want to make money with your emails. So, focus on subscribers who you can turn into customers in the future. Make your emails compatible with latest mobile technologies to become a vital source.

When generating freebies for subscribers, don't give everyone the same option. Some might like a checklist, while others will look for an eBook.

To engage subscribers, understand their buying intention. Then, create a small piece of a product as a free model. This will help your subscribers and make them more interested in your paid products. Remember, freebies are not to get more subscribers. They are provided to lead towards product pitch.

Here is a summary of every step you need to accomplish as an email marketer:

- Spend hours researching your target market to know their likes and dislikes.
- Use the obtained conclusions to define product needs that you can fulfill.
- Leverage multiple techniques and tools to build your email list. Focus on getting a targeted list that will prove useful for your product or service idea.
- Learn about segmentation and gather resources to start segmenting. Choose segments that are more valuable to campaigns you have in mind. You can even define multiple segments for different products you have.
- Improve your email list by qualifying every subscriber. Make sure every subscriber knows and wants your services.

- Design a value providing email sequence for different segments. Constantly stay in touch with your subscribers via emails that offer quality resources.

- Choose different methods to sell your products or affiliate products. Use long-term techniques to keep on making money from your emails.

- Follow an automated sequence that works on its own, so that, you can keep on passively getting revenue.

- Learn all about email marketing metrics and gather reports through analytics. Keep monitoring open rates, conversions, and other metrics regularly. Modify your email campaigns according to the available data and maximize ROI.

All in all, email list building and making money depends on how valuable you are in the eyes of your subscribers. The more you offer, the better you become.

So, start building your email list and invest time in gathering effective marketing tools. The initial phase requires dedication and patience. But then, you will create an automated cycle of passive income.

Share your knowledge with others and help them earn as well!

References

http://www.marketing-schools.org/types-of-marketing/email-marketing.html

https://www.roimachines.com/10-email-marketing-benefits/

https://www.pure360.com/10-benefits-of-email-marketing/

https://www.entrepreneur.com/article/285949

https://www.verticalresponse.com/blog/40-brilliant-but-easy-ways-to-build-your-email-list/

https://socialtriggers.com/list-building/

https://www.dailyblogtips.com/how-to-build-an-email-list-that-makes-money/

https://optinmonster.com/50-smart-ways-to-segment-your-email-list/

https://blogs.constantcontact.com/profitable-email-marketing/

http://traceylawton.com/blog/build-your-relationship-with-your-subscribers-with-these-7-simple-ways/

https://blog.leadquizzes.com/12-clever-ways-to-make-more-money-from-your-email-list

https://wppopupmaker.com/email-marketing/make-money-with-your-email-list/

https://www.melyssagriffin.com/make-money-email-list/

https://help.activecampaign.com/hc/en-us/articles/206907550-What-is-an-opt-in-list

www.ingramcontent.com/pod-product-compliance
Lightning Source LLC
Chambersburg PA
CBHW071207210326
41597CB00016B/1707